Presented to

Aileen Aderholdt, Librarian
for

Carl A. Rudisill Library
of

Lenoir Rhyne College
by the author

*E.M. Coulter*

September 3
1973

# DANIEL LEE
# AGRICULTURIST

*His Life North and South*

DR. DANIEL LEE

# DANIEL LEE
# AGRICULTURIST

*His Life North and South*

❧ ❧

By E. MERTON COULTER

UNIVERSITY OF GEORGIA PRESS • ATHENS

# ❧ Contents ❧

# ❧ Preface ❧

DANIEL LEE was one of the minor figures in American history, but a major leader in the field of agriculture who strangely has been neglected for three quarters of a century. His life and career were significant in several respects. He was born in the North and lived about half of his life there before moving to the South where he spent the remainder of a long life of almost ninety years. His span of years covered that most interesting and tragic period of the Civil War and Reconstruction which was shot through with sectional animosities. It was followed by an attempted reconciliation which had not completely succeeded a hundred years later.

He was born in the state of New York, where he lived until he moved to Georgia in 1847. During the Civil War he moved to Tennessee and died there in 1890. He considered himself an American rather than a Northerner or Southerner. But he held principles from the beginning which were more in keeping with those of the South, and, therefore, he readily accommodated himself to life there, although he was accused by some Northerners of having deserted them in mind, and, indeed, as he had done in body.

As an agriculturist, Lee had fixed convictions on what was best to do. He was first of all a soils chemist (and a practical geologist), who felt that depletion of the fertility of the soil if not checked would finally result in the destruction of the republic. Therefore, high in his agricultural program were the prevention of depletion and restoration of fertility of the soil, both by proper farming methods and by the use of organic and chemical fertilizers. Diversification by the rotation of the old customary crops as well as by the introduction of new ones played their part in what he advocated. He was almost hypnotized by the many grasses which he studied and whose introduction he recommended in season and out. At times he emphasized livestock and grass economy almost to the exclusion of the plow, especially after he moved to Tennessee and during the time of the Reconstruction when the freedmen had not yet adjusted themselves to the necessity of settling down and be-

vii

coming dependable laborers, willing to earn their living by the sweat of their brows.

To promote the various elements in his farm program (including the gathering of agricultural statistics), Lee became an inveterate editor of and contributor to farm journals. Even before going to Rochester, New York, where he spent the most important part of his life in the North as editor of the prestigious *Genesee Farmer*, he touched on agricultural subjects in his contributions to a Buffalo newspaper, but his fame as an agriculturist in the North rested on his editorship (and ownership for a time) of the *Genesee Farmer*. After moving to the South, he edited for more than a dozen years the *Southern Cultivator* in Augusta, Georgia. Thereafter he edited the whole journal or the agricultural section of such ones as the *Southern Field and Fireside* (Augusta), the *Plantation* (Atlanta), the *Farmer and Artisan* (Athens), and the *Tennessee Farmer* (Nashville); and to various other farm journals he contributed articles.

In addition, as head of the agricultural section of the Patent Office in Washington, he carried forward his program in the Patent Office reports. Facile equally with tongue and pen, he had the distinction of being elected in 1854 the first Professor of Agriculture at the University of Georgia, indeed the first or among the first in any American college or university. He remained in this position a half-dozen years until the University was interrupted by the Civil War. Here in the classroom he preached verbally to the sons of planters who had been reading his sermons in the *Southern Cultivator*.

Lee's traveling back and forth between the North and the South while he edited both the *Genesee Farmer* and the *Southern Cultivator* for a period of years, his residing for short spells in Washington while connected with the Patent Office, his moving about within the South, his going to Knoxville and then to Nashville, and his losing two residences in Tennessee by fire—all of this mobility prevented him from accumulating and especially from preserving what would have been a most extensive and valuable archive of personal correspondence and an excellent agricultural library.

Therefore, this biography had to be written almost entirely from printed sources, the main exceptions being the records of the University of Georgia and a collection or two in Washington relating to the Patent Office. As might be expected, the most valuable sources were the *Genesee Farmer*, the *Southern Cultivator*, and various newspapers. The bibliography attached to this work gives a complete listing. It should be said here, that apart from the *Gen-*

*esee Farmer* and a few other Northern farm journals and news-papers, the records of Lee's life in the North are scarce. It should not seem illogical, therefore, that the greater part of this biography relates to Lee in the South, where the most significant part of his life was spent.

It is evident that Lee was out of tune with twentieth-century thought on slavery and the Negro; but the purpose here has not been to interpret history apart from the times when events oc-curred, but rather to record them in the spirit of those times. It would be writing another book to hold Lee up before a mirror of the late twentieth century—much like blaming Julius Caesar for riding in a chariot instead of in an automobile.

E. M. C.

# ❧ I ❧

# In New York

DANIEL LEE had a career almost equally divided between the North and the South. Without letting geography influence the programs he advocated or his principles or his philosophy of life, he attained national prominence in both sections of the nation. Yet after a long life he died almost entirely unnoticed—"unwept, unhonored, and unsung."

A strange obscurity pursued him after his death, for no complete sketch of his life seems ever to have been written.[1] Had he not used a literary device in his writings of often reminiscing incidentally in making a comparison or striking home a point, much of what is known about him now would never have survived.

He was born in Herkimer County, that elongated political division of central New York, in 1802 (or possibly in 1808), on his father's farm.[2] He became thoroughly imbued with farm life and its problems, and although he developed other interests, his life's work revolved around agriculture. As a boy he worked on the farm; he plowed, hoed, mowed and put up hay. He looked after a small dairy, and he helped manage other livestock, later recalling that he grew up "on a pretty large stock farm."[3] In 1828, after he had left the old home, he liked to remember that his father had put up a large stack of hay and had called it his "Jackson stack" in honor of Andrew Jackson, who had been elected president that year.[4]

In his work on his father's farm young Lee had the help of other members of the family as well as of a few slaves, which his father owned. Slavery was an ancient institution in New York, having been brought in during the early days of the colony, and was to continue in slight vestiges until 1841.[5] Lee was not much impressed by the industry of the slaves, and much less so after they had been freed. In his later life, he said, "I remember well how worthless the young fry became after emancipation." They tried out as hostlers, barbers, bootblacks, and house servants, in all of which occupations they improved their lot little or none.[6]

1

Lee had been largely self-educated, certainly beyond the common schools of the neighborhood. As he once remarked, he "learned his Greek Grammar while the oxen with which he plowed were dining on hay or grass."[7] Impressed then as he was for the rest of his life on the value of grasses, several varieties of which grew on his father's farm, he gained the reputation of being "one who had acquired his education by making two blades of grass grow where only one had grown before."[8]

As a young man, he added to his agricultural interests the study of medicine. In the nearby town of Fairfield he attended "medical lectures" at Fairfield Medical College, which was allowed to award a doctor of medicine degree. Because of Lee's receiving this distinction, throughout the rest of his life he was known as Dr. Lee. Always identified with western New York before going South, he began the practice of medicine in Chautauqua County, the westernmost county in the state. Soon he moved into the adjoining county of Erie with its bustling city of Buffalo, where he located. If he practiced medicine here, it was only incidental to other interests which he developed. It must have been about this time when he made a visit to the Midwest. As he put it, "Long before I moved to Rochester, I spent over a year up the Valley of the Mississippi, studying its farming operations, and agricultural resources."[9]

In Buffalo and Erie County, Lee soon attained a position of leadership in agricultural, educational, and literary activities. Intrinsically an agriculturist, he could not rest satisfied until he acquired farmlands and stocked them with cows, which he bought at $12 a head.[10] For a dozen years and more he was a school official (superintendent or inspector of the common schools), and when President Jackson distributed the surplus revenues, Lee became one of the commissioners to manage the part allocated to Erie County.[11]

Lee's most continuous activity in Erie County was in the field of agricultural journalism, a field in which he had first begun writing articles for newspapers and journals when he was only twenty years old. And now for a time, in Buffalo, he published a paper called *Honest Industry*, which was devoted to agriculture and industrial developments. He later said that he published the paper "to cultivate and elevate the industrial mind of the United States." In editing and publishing this paper, he collected statistics on agriculture, manufacturing, and mining. Knowing that Henry Clay would be much interested in this information, Lee sent him copies, for which Clay duly thanked him. Lee said that this pioneering work led to the inclusion of schedules on such statistics in the United States census of 1840.[12]

In the fall of 1838, Lee ventured into a part ownership of the *Buffalo Journal*, which he held until May, 1839, when this paper was merged with the Buffalo *Commercial Advertiser*. For this paper Lee wrote a series of articles on agriculture and continued to write occasional letters for sometime afterward.[13] Also, he wrote for other newspapers and journals during this period and thereafter, including the *National Intelligencer* in Washington.[14] And so he became acquainted, either personally or through correspondence, with many of the agricultural editors throughout the nation, one of the best known being John S. Skinner, who was founder of the *American Farmer* and was "known to the writer for many years," as Lee put it.[15]

Lee, as facile with his tongue as with his pen, never let either one stay still for long. He said in his later life that he had lectured free of charge in "nearly every county" in New York, and he might well have added (for it became a matter of record) that he lectured as far afield as Boston, Massachusetts.[16]

These lectures were not only parts of formal programs of agricultural societies, such as the New York Agricultural Society, the Rensselaer Agricultural Society, and the Erie Agricultural Society, but also before local groups. Lee always carried the message of agricultural betterment, which burned so glowingly in his heart. For some years he was the corresponding secretary of the New York Agricultural Society.[17]

Lee had gone far enough along as an agricultural expert while still living in Buffalo to lead to his election to the New York legislature in 1844, where in the House he was appointed chairman of the Committee on Agriculture. He was a "worthy representative of the agricultural interests of the farmers of Erie and the state," Luther Tucker, editor and proprietor of the *Albany Cultivator*, said.[18] Lee immediately addressed himself to what was heaviest on his heart, the welfare of the New York farmers.[19] He framed the bill, which became law, requiring the collection of statistics every ten years on the number of acres planted in various crops, the yield per acre, and on other subjects. Lee's main program was to promote in the legislature aid to the farmers and laboring classes, and this he hoped could be obtained through an appropriation for an agricultural school and for lectures on agricultural subjects to be given widespread over the state—those who could not come to the contemplated agricultural school would be served by agricultural information being taken to them. The high position Lee had reached as an agriculturist in the eyes of New Yorkers easily suggested his appointment to the Committee on Agriculture and later his appointment as chairman.

The report he wrote became celebrated, both to be highly praised and to be assailed in a "storm of criticism." Having come along through hard work, Lee was to confess "that all my sympathies are with the laboring classes."[20] His mission in life was to help the farmers and producers. "For these people I feel a profound sympathy," he said. "Most of them are alike distinguished for their honesty and patient industry. They produce much, because they are always at work; they consume little, live poor, and die poor, to enrich others." It pained Lee to "reflect on the fact that the number of public paupers in this great and wealthy state increases much faster than population."[21] Or as he put it later, "It is alike unsafe and unnecessary, to manufacture paupers, and criminals, a great deal faster than our own population increases."[22]

Lee took advantage in his report to argue the cause of the poor. With unusual sharpness and clarity he laid down his philosophy of a proper social organization, one based on as even a distribution of property as the rewards of labor should dictate, whether the laborer be a field hand, a worker on the Erie Canal, a woodchopper in the forests, a seller of goods, a teacher in schools, a dispenser of medicines, an arguer before the courts of law, or whoever. As he saw the situation in New York, the rich were getting richer and the poor poorer; because the day laborer (and that meant generally, in Lee's thinking, a farmer) was being cheated out of the just rewards of his labor by a system that allowed merchants, physicians, lawyers, and others in the upper professions to receive for their day's work vastly more than the lowly worker got. "If a man whose whole life is devoted to the cultivation of the earth," said Lee, "does not and cannot earn so much as the merchant, the physician, or a lawyer, in the course of a year, pray tell us what is the *cause* of this inability. . . . The hard work of skillful farmers is bought and sold at 9 or 10 dollars a month, and twelve hours toil is cheerfully performed each day. But the mechanic, the banker, the merchant, the broker, or the professional gentleman, thinks his service very poorly rewarded if he does not receive three or four times that sum."

Continuing, Lee said that it was a notorious fact that "the great body of our rural population somehow contrive to work a little harder and fare a little poorer than any other class of the community." On what did the productiveness of the farmer's labor rest, Lee inquired—surely not on his muscular strength alone, "for in that case the mechanical power of a cart-horse will exceed five-fold in value the labor of an agriculturist." It all went back to educating the farmer in his skills: "There is no power on earth so productive of great and beneficial results as the power of highly cultivated intellect." Here was the answer of what Lee was driving at—an agricultural school to be set up by the state, where ignorance would be routed and scientific knowledge enthroned. It was true, he said, "that science is the greatest leveler in the world; but, unlike the leveling of ignorance and brute force, it ever levels upward."[23]

Some of the state papers were highly indignant at this report and accused Lee of trying to set one class against another. He was accused of being an agrarian leveler, but as Karl Marx's *Communist Manifesto* had not yet been written and Marxism was unknown to the ordinary citizen and newsman, Marx did not get into the dispute. One New Yorker, in defending Lee, said, "Strange it is, that the moment an individual has the philanthropic boldness to advocate the rights of the producing classes, the whole kennel of those who hope to live by their wits, are out upon him, 'tray, blanch, and sweetheart.' "[24]

One of Lee's critics, whom he chose to answer in his *Genesee Farmer*, was the editor of the *Ithaca Chronicle*. "He knows little of human nature," Lee said, "who supposes that a majority of the free citizens of this State will be content to work hard, and fare hard, live miserably and die, either in, or out of a poor-house, to enrich others."[25] Lee recalled that when he was eighteen the most he could get for cutting 100 cords of hardwood and board himself was $30. What was the justice in a system, Lee wanted to know, that would pay *Farmer* Lee only $30 for cutting 100 cords of wood, but would pay *Doctor* Lee "an equal sum for cutting off a leg about as easily as he could fell a basswood sapling."[26] The remedy for this "enormous evil" of such discrepancy in wages was not "in arraying one class against another—the poor against the rich—but in a common and generous effort so to enlighten the popular mind" in all matters relating to their lives. Crime would then vanish or be greatly lessened, and so would discontent, poverty, and the most of taking care of the poor.[27]

It might be suspected that Lee's social philosophy as expressed in his report to the New York legislature and in his defense against his critics would set him down as a somewhat emotional reformer for the rest of his life; but this was not to be, for his enthusiasm was always directed primarily to the principles of agriculture and agricultural reforms, not social reforms. Yet for the next few years he was not to forget to stand up for laborers and for what might be termed the rights of the "little man." "If you believe," he said, "that you are not a brute—you should cherish some hope of Heaven, some fear of a just God—then read your Bible, and believe it when it tells you that, 'it is easier for a camel to pass through the eye of a needle, than for a rich man to enter the kingdom of Heaven.' " He did not like an unregulated capitalistic system that made the rich richer and the poor poorer; but he wanted it to be made clear that he believed that "agrarianism, and all ideas of a division of property, are at best mere quack remedies, calculated to do infinite harm rather than good."[28]

In standing up for the laboring man and the poor, Lee was equally opposed to their hoping to get something for nothing. Man should live by the sweat of his brow. Labor was elevating, not degrading. "Christian civilization" should impel "young, and old, and middle-aged persons, of both sexes . . . [to] work steadily and successfully from an inborn desire to do good in the world, after every mere animal want is satisfied. The millions work to live; and when any can live without work, they do so as naturally as a herd of buffaloes on the western plains."[29] He also said, "Mere negative existence, a life of idleness and emptiness, will not excuse moral and intellectual beings that were created to labor alike for their own good and the highest happiness of the race."[30] Lee abhorred the drones who abhorred labor: "Labor is the grand humanizer of our race. Without it, man never rises but a single step above the speechless beasts of the field. With it each generation may excel all preceding ones to the end of time."[31] Lee would have been greatly shocked at the twentieth-century government welfare system.

But Lee, now in the New York legislature in the session of 1844–1845, was convinced that a proper regard by the state government for all the people alike by providing all with education and training would solve every social and economic ill. Occupying the great void in that program were the farmers and other laborers. Now it was time to quit talking about setting up an agricultural school, which had been going on intermittently for almost a quarter century. It was now time to start one, and to that end, Lee introduced a bill in the legislature to appropriate to the Fairfield Medical College (Lee's alma mater) $5,000 for each of three years if it would change into an agricultural school. Its buildings, chemical apparatus, other facilities, and its location in a small country town (in southern Herkimer County) would make it well fitted for an agricultural school. Nothing came of this, but Lee's efforts bore fruit three years later when the state legislature did charter such a school. Lee's idea was that with the success of the Fairfield agricultural school, others would follow—"with the ice once broken in the matter of agricultural schools, and their practical utility fairly tested, there will be no difficulty in establishing them wherever they should be needed."[32]

Not to be stopped by the refusal of the state to act, Lee associated himself with his old friend Gen. Rawson Harmon, a practical farmer who lived about twenty miles from Rochester on a 200-acre farm near Wheatland in the southwestern corner of Monroe County, to set up an agricultural school on his farm. The students were to live in Harmon's home and take their meals there, or if the number of students became too large, then they could be accommodated in nearby farmers' homes. They were to study agriculture both in the fields and in the lecture and laboratory rooms. Lee would give lectures on agricultural chemistry and conduct experiments in the classroom and in the fields. General Harmon would be the principal of the school and do teaching as required. A cultured gentleman of liberal education would teach the ordinary branches of learning. There were to be four terms at $25 dollars tuition each or $100 for the full year. There would be also a special short term of a month in the fall, conducted by Lee, for schoolteachers to acquaint them with the fundamentals of agriculture and the chemistry of soils. The school was to be set going in May, 1846. It was given the high-sounding name of the Western New York Agricultural School.[33]

The school was opened according to schedule, and during the first year fifteen or sixteen students attended. Lee had for some time toyed with the idea of writing a textbook on agriculture, but the nearest he seems ever to have come to doing so would have been a collection, never undertaken, of his vast writings on the subject. But now, as a help to beginners in agriculture, he published in the *Genesee Farmer* an article entitled, "The Study of Agriculture. Chapter I," but the promise in this title of other chapters to follow was not fulfilled. The article was in question-and-answer form: "WHAT is agriculture? 1. Agriculture is both an ART and a SCIENCE. What is the *art* of agriculture? 2. The *art* of agriculture consists in the skillful application of manual labor to the cultivation of the earth. What is the *science* of agriculture? 3. The *science* of agriculture has for its object the investigation of the Natural Laws and changes in Matter, by which it is *organized* into the bodies and products of cultivated plants, and domestic animals." And so it went through forty-two questions and answers.[34] At least on one occasion Lee found time to take a group of his students on a trip from Lake Ontario up the Genesee River to examine the rocks and study the geology of the region.[35]

Quite enthusiastic in the beginning, Lee set about raising a fund sufficient for erecting buildings to accommodate a hundred students; but as too often happened, his enthusiasm ran away with him and he got nowhere with his campaign. He soon decided that it would be better to have the school located near a big city, and Rochester being such, he hit upon that place for this reason and for another very important one. In 1844–1845, at the very time when he was serving in the New York legislature, he made arrangements to become the editor of the *Genesee Farmer*, the outstanding New York agricultural journal which was published in Rochester. He continued for some time to consider Buffalo his home, but he was in Rochester as time permitted and to get out his first issue of the *Genesee Farmer* (for February, 1845). Later in the year after the legislature had adjourned and during the following year, he was in Wheatland to organize the school and to lecture in it. But considering the twenty-mile trip down to Wheatland too time-consuming and otherwise inconvenient, he planned to get the school moved to Rose Hill, just outside Rochester. Here in the city were the *Genesee Farmer* editorial offices and the plant where the *Farmer* was printed. A half-dozen years later Lee would be part owner of this printing establishment.[36]

Famous in the region and beyond, for that matter, as a practical and scientific farmer and a bold legislator, Lee was enthusiastically received in his new position by the agricultural press of the country. A modern historian has referred to him in this connection as a "nationally known agricultural lecturer and one of the best agricultural chemists and educators of the period."[37] Always a practical, as well as a scientific, agriculturist, Lee could not resist engaging in on-the-ground agricultural activities—here by becoming associated with a dairy farm consisting of fifty cows.[38] The move to Rose Hill did not take place.

Lee was a man of destiny in a minor way; he moved far and wide geographically, always pursuing his first love of helping the farmer and the laborer. In the midst of his editorial work and his planning for moving the agricultural school, he received a call from the South to carry on his work there. And without giving up at this time his editorship of the *Genesee Farmer*, he left New York, never during the remaining forty-three years of his life to return as a permanent resident. In the South he would expand and carry forward the preachments that had made him eminent in agriculture. He was now to become a Southerner of Southerners, never to abandon his love and respect for his native state. But for some years thereafter he would be back occasionally in Rochester to attend to editorial duties there, and at one time he had almost made up his mind to reside there permanently.

# ❧ II ❧

# Lee Moves to Georgia—Lee the Man

LEE'S CAREER in the North had not gone unnoticed in the South where his program of agricultural betterment and especially his desire to set up schools to teach agriculture appealed much to Southerners who were interested in promoting the prosperity of their section. At this time, one of the best-known farm magazines in the South was the *Southern Cultivator*, founded in Augusta, Georgia, in 1843. Two years later, James Camak of Athens became its editor; in June of 1847 he died, after having given the *Cultivator* the high standing which it enjoyed.

The proprietors, James W. and William S. Jones, immediately set about looking for a new editor with a reputation sufficient to maintain and even improve the excellence which Camak had given it. Daniel Lee of the *Genesee Farmer* came to mind, for already he had written several articles for the *Southern Cultivator*, which also had reprinted a few from the *Genesee Farmer*. Negotiations were immediately started, and soon the proprietors were able to announce that they had secured Lee's services "at a great expense." In making their announcement, they referred to him as "among the most scientific and practical Farmers of the Union."[1] They also stated that he was the author "of many valuable essays on Agriculture, and its kindred sciences." He was favorably known throughout the Union "as a ripe scholar and practical agriculturist . . . [having] a mind highly cultivated and enriched by the acquisition of much and valuable information, obtained in the wide field of practical experiments, to which much of his time and labor have been devoted."[2]

Lee assumed his editorial duties with the August issue of 1847. Being a Northerner now in the South for what he doubtless considered a permanent residence, he was highly regardful of the proprieties in such a situation, and he so expressed his feelings in his salutatory. He hoped that his Southern association would "ripen into mutual esteem and lasting friendship." He was sensible of the

10

fact that he might commit errors because of his "ignorance of Southern agriculture, climate and soils; and of the habits, and customs of the people so differently situated from those of the Northern States." He asked the readers' indulgence, which he felt would be freely given.[3] Knowing that politics was a deadly game, probably more so in the South than in the North, Lee was quick to state that the *Cultivator* "must be free from every thing of a political complexion, but exempt from the *suspicion*, that any party or sinister purpose whatever, lurks about it."[4] In noting Lee in his new position, the editor of the *Prairie Farmer* of Chicago said, "Dr. Lee is an able editor; and though fond of riding with a vengeance, when he strides a hobby, he does know how to make a good paper."[5]

While contemplating moving his family from Rochester to Augusta, Lee led what he termed "an office-life," which freed him to come and go over the state to get acquainted with the people and their problems and ideas, and with the lay of the land and its geology and resources. Already acquainted by reputation with such leaders as Richard Peters of agricultural and livestock reputation and with David Dickson, the well-known planter who experimented with fertilizers and cotton, he would soon be meeting them in person. In mentioning Dickson, Lee said that this Hancock County farmer had done more than anyone else to put Southern farming on an enduring basis.[6]

It seems that Lee's family, consisting at this time of his wife Sabrina and several children, never moved to Georgia. Therefore, to visit his family and to look after other interests in Rochester, he made frequent trips back and forth. His other interests consisted in editing the *Genesee Farmer*, which he did not give up for some years, and from 1850 to 1856 in looking after his association with the *Rochester Daily American* newspaper. For the first few years he spent the winters in Georgia and parts of the summers in New York. Some summers he spent mostly in the South, traveling widely over Georgia and elsewhere.[7]

Lee was fascinated by travel; he was almost a gypsy. Traveling for a purpose besides getting there, he noted agricultural conditions along the way, a description of which he would likely publish in one of his two journals or in some other place. He advocated travel as good for people and especially good for perpetuating the nation. "Travel will cure a thousand prejudices and errors which every man unconsciously falls into," he said. Furthermore, "Cheap travelling and greater intercourse are necessary to remove sectional prejudices, and make us one in feeling and sentiment, as we are one in language, religion, government, and interest."[8] To facilitate travel, Lee favored a rapid development of railroads and steamboats, which would "mingle the citizens of every State with those of all the others."[9]

On almost any day he could take long walks up and down the Savannah River from Augusta, "studying rocks, minerals, soils, plants and animals." And as South Carolina lay across the river, he did not neglect the Palmetto State.[10] He wandered widely over Georgia from Rabun County on the North Carolina line to Chatham on the coast, getting down to the "grass roots," spending the nights in farmers' homes and making friends with their families. Although at times partaking of the character of a footpad, also, he traveled on much higher planes, as he attended county fairs and county and state agricultural society meetings; often he lectured on such occasions. The first agricultural address he made after arriving in Georgia was in Sparta in Hancock County, the home of William Terrell, an outstanding planter to whom he would later be grateful and whose name he could never mention without adding encomiums of affection and praise.[11] He gave public lectures in Athens, Watkinsville, Marietta, and in other towns where he talked to full houses. After visiting Jefferson County which adjoins Richmond County, he declared it to be "one of the best farming counties in Georgia." Having been a member of the New York legislature, he visited the Georgia General Assembly on occasions when he sought its aid in promoting certain of his agricultural interests.

Not only did he carry on his walks across the Savannah, but now and then he lectured in South Carolina towns, such as Abbeville and Calhoun Falls.[12] And the year of his arrival in Georgia (1847), he attended the commencement exercises of South Carolina College (later named the University of South Carolina) in Columbia and was much impressed with the financial support the state government was giving the institution.[13] Lee had no inhibitions in approaching strangers or being approached by them. He was always ready to engage in a pleasant and friendly conversation. A South Carolinian who had met him on a train and later had occasion to write him remarked about "the very pleasant acquaintance and conversation with you which accident procured me, on the steam cars."[14]

Lee was never wanting for words to clearly express his thoughts or to turn an apt phrase. In a light and somewhat humorous vein he would refer to a person as "a two legged talking animal."[15] There were expressions he frequently repeated for emphasis, for instance, "as long as grass grows or water runs" to denote a long time or eternity.[16] Lee's style led other editors through carelessness or downright plagiarism to copy him without giving him credit. Since in this way his thoughts were being widely spread and since his main purpose in life was to promote agriculture, he did not complain much. Yet he could not help but say, "We have suffered depredations of this kind so long, that we very rarely provoke a passing remark."[17] Or in another instance, "Although none have suffered more by the pilfering propensities of their contemporaries, none have complained less." He declared that he hardly ever opened and looked at his exchanges without some of his writings being copied without credit being given to him, or what was worse, credit given to some other publication which had copied Lee without credit.[18] Despite his clarity of expression, Lee had a habit of switching without proper transition to something quite off the track of his main topic.

Although it is not known that Lee belonged to any organized religion, he was a firm and devout believer in a Supreme Being, as he so often showed in using such expressions as the "Divine Hand," the "great Creator," or "God's Laws" when describing the forces of nature—what made plants grow, water run downhill, or the wind blow. No evidence has been found to indicate that he might have been a Quaker, despite his occasional use of "thee" and "thou" and such expressions as "didst thou."[19] He saw all nature following a divine pattern. The profession of the farmer embraced "the most sublime truths in the mineral, vegetable, and animal kingdoms. These kingdoms, in the providence of God, meet, mingle and blend harmoniously together, on the surface of our planet."[20] To carry out this idea in the language of the poet Alexander Pope, Lee quoted his couplet:

> All are but parts of one stupendous whole,
> Whose body Nature is, and God the soul.[21]

And Lee, if knowing and choosing to quote Shakespeare in *As You Like It*, also could have expressed what was in his mind in this fashion:

> And this our life, exempt from public haunt,
> Finds tongues in trees, books in the running brooks,
> Sermons in stones, and good in every thing.

Lee did say, referring to his scientific investigations, "We now see beauty and design where before we saw no beauty—nothing to attract our attention. We now hold converse with nature, for we have learned its language. The rocks, the trees, the insects, speak to us of their age, construction, and destiny. The broken rock that once we would have passed unnoticed, is now examined; and it affords food for thought—we trace it in the leaf or the insect, and thus learn that once it belonged to the vegetable or animal world. Science teaches that this rock is crumbling and forming soil—that it contains elements that assist in forming bread and supporting life."[22] Lee was, indeed, much interested in geology.

In a special sense Lee was a philosopher, at least an agricultural and nature philosopher. As an editor announcing a new year, he would go into a long exercise of offering good advice and calling for respect for the Deity, and occasionally he would recall axioms and articles of faith, which he hoped his readers would accept and embrace concerning the problems of life, the proper order for an economic society, and so on.[23] In 1847, he went through this little philosophical exercise: "What is *Time?* Our largest idea of it gives but a point on that vast and that incomprehensible circle designated by the name Eternity. . . . The Progress of Man in Knowledge, in Virtue, and Philanthropy, is capable of being reduced to a science. Divine Revelation discloses the true basis on which this science must be erected. Man must evince his sincerity of his love to his Maker and final Judge, by showing in his daily conduct that he truly 'loves his neighbor as himself.' Pretension will never answer."[24]

Included in love for the Maker of the great universe were love of country and thanks to the Supreme Being, who had so richly endowed it. Bespeaking this love which his countrymen had, he said, "Heaven has kindled a vestal fire on our hills and mountains. It has appointed, not Roman Virgins, but iron-palmed husbandmen, to protect and keep alive an ever growing and sacred flame. . . . Our agricultural, mineral, manufacturing, and commercial resources, are altogether beyond computation. The danger is, that we shall prove unworthy of blessings, so numerous and transcendent. Unmerited wealth and undeserved prosperity have ruined millions of individuals, and induced the speedy downfall of many powerful nations."[25]

With such a mind at work, it would be surprising if it did not wander off into the historic past. He often called upon his knowledge of history from the dawn of civilization to the present. In describing the development of the plow, he began with the early Greeks from Hesiod and Homer and continued to the modern era.[26]

He liked to use the old axiom, attributed to Dionysius of Halicarnassus, "History is philosophy teaching by examples."

Although Lee never threw New York or his life there into the faces of his Southern readers in an unbecoming way, he could scarcely write about anything without making some comparison with the Empire State of the North—his now being in the Empire State of the South. As often as not the comparison was to the disadvantage of his native state, and always so when he was discussing Southern and Northern climates.

Lee truly believed that it was his divine mission to improve the lives of the people, principally, of course, by availing themselves of the better methods of farming and by taking every advantage of what nature provided. Yet he was not a flabby sentimentalist who talked and wrote as if he were everything to everybody. He had fixed principles and he would defend them to any extent short of violence; on at least one occasion he intimated violence in defending his good name. He was not a person to be cried down by charlatans or taken in by humbugs; although he was not a vain person, he stood up for what he had done or discoveries he had made. Sometimes when it was only a matter of what chemicals were in a certain soil or how much a certain fertilizer would hold of certain chemicals, he could be quite caustic in his disagreements. So it was that when some absurd statements which a Prof. Charles Baer made in a lecture in Sparta were told to Lee, who was not there, he firmly disagreed and said that Baer was one of those phony professors who assumed the title to enhance his standing. When Baer convinced him that he was a genuine professor in a Maryland college, Lee apologized, but still disagreed with him on some of the professor's chemical statements.[27] As an example of Lee's upholding his priority of discovery, he showed conclusively that he had antedated a person who asserted the claim of having discovered a potato plant disease. Lee admitted that he had not found the remedy.[28]

Another characteristic of Lee should be emphasized. In coming South and embracing Southern ways of life and thinking, he was in no way guilty in going back on his former Northern breeding and principles. He was never a sectionalist. He was never tinged with abolitionism, and he knew the institution of slavery as it existed in New York. He would learn much more about it in the South and would vigorously defend it. When he came South, he gave up nothing except eventually his property and residence in the North. There was no trace in him of toadyism or a patronizing air toward the South.

In his divinely appointed mission to improve the lives and fortunes of country people (and all others incidentally), Lee was
merely to carry on and intensify what he had been doing in New
York. But let it be said that at times he placed the enjoyment of life
as something above the mere fact of making a living and accumulating money and property; though, of course, Lee's lifework was promoting the latter, which, after all, would make possible the former.
In supporting the enrichment of country life, he wrote, "Landscape
gardening, groves, orchards, lawns, fountains, and whatever can add
beauty or interest to a rural residence, will be regarded as deserving
care and study. To sacrifice one's whole life to bags of cotton or bags
of gold, is a manifestation of idolatry, such as many a pagan would
be ashamed to display." [29]

Now and then Lee wrote articles in a series entitled "The Improvement of the People" in which he set forth the many facets of
his program for agricultural reforms. [30] Naturally he advocated and
supported agricultural societies. There were a small number of
local ones in Georgia, and a regional one, organized in 1846, was
entitled the South Central Agricultural Society. Soon after coming
to Georgia, Lee was elected its corresponding secretary, a position
which he held for a term. [31] As good as these societies were, still there
was something lacking, for they did not reach the people in their
home communities, where individual planters might meet together
in clubs and discuss their problems. These small groups could set
up little libraries of agricultural books and journals, which might
be placed in a planter's home, or they might be taken care of in a
special building in some small town. Lee observed, "No other com-
modity costs a man so much through life as his needless ignorance." [32] Thus could people escape being imposed upon by the
purveyors of humbugs, so flourishing in those times. "The people
buy humbugs because they love them. If they did not love them,
they would not patronize them so extensively, nor receive their
nostrums as a free gift. True science is no humbug, and, therefore,
it is not sought after by the millions." [33]

Lee looked not only to the organization of smaller groups to improve the tillers of the soil, he also envisaged the organization of all
the "friends of agriculture in a common effort for its elevation at
the South, since it is plain to our mind that its profits and honors
would soon be doubled what they now are." [34] He asked the question, "What is the repelling force in Southern society that a thousand planters can not possibly act harmoniously together to advance
their noble calling?" He said that they knew they had great agricul-

tural resources "and yet all do next to nothing, in concert with others, to improve their condition." [35]

One answer to this situation, Lee thought, was to educate the people in their schools to recognize the problems and to know the solutions to the problems encountered in farming and planting. This was much on his mind, and he frequently wrote about it. It had come to be customary to lay the blame for the impoverishment of Southern soil on the slave system of labor, but he declared that after long and careful study in both North and South, he had become satisfied "that the relations subsisting between employers and employed in any quarter of the Union had little or nothing to do with it. Being mainly a scientific problem, the popular understanding, both North and South, will fail to solve it" and would never solve it "so long as the scentific principles of agriculture are disregarded in the common school education of the sons and daughters of American farmers." One looked in vain for a single textbook "in our most cherished institutions of learning" which was designed "to teach the true balance of Organic Nature, between cultivated plants and cultivated animals, and between both and the land that supports them." None of these things commanded the patronage of Congress or a single legislature. Writing this in 1855 he concluded, "Not one of these bodies has ever given a dollar, to our knowledge, to found an agricultural school of any kind." [36] Lee, who considered himself a scientific, as well as a philosophical, agriculturist, asked, "If science has done so much in its feeble infancy, what may we not expect, when every citizen may be more educated and better informed than the wisest now are?" [37]

In 1850, Lee was calling for Congress to give 100,000 acres of public lands to every state that would appropriate $50,000 to establish a normal agricultural school designed to educate teachers for common schools and academies "in all the natural sciences, which serve to develop and illustrate the principles of agriculture." [38] This system would reach a great many students who would never go on to colleges of agriculture—*colleges of agriculture,* not simply the old established literary institutions with a chair of agriculture tacked on. No college of agriculture (and none had yet been set up) should have fewer than six professors to teach the various branches of the subject, for otherwise it would be like having one professor to make up the faculty of a medical school. Added to a college of agriculture there should be an experimental farm of several hundred acres. [39]

About a half-dozen years later in 1857, Lee was congratulating the state of Michigan on being "the first to found and put into successful operation a State Agricultural College in the New World."[40] After two more years, he had the pleasure of noting that the Maryland Agricultural College would soon go into operation, and with equal, if not more, pleasure he learned that the cornerstone of the New York State Agricultural College had been laid. This fact led him to reminisce that it was back in 1822, when he was on a visit to Albany, that a bill to establish an agricultural college had been introduced in the legislature and killed. He added, "But from that day to the present, the plan of having such an institution has never been abandoned; and after the lapse of thirty-seven years, and half as many defeats," the college was now on the way. But a "whole generation of old fogies has passed off the stage for their country's good, while the friends of agricultural learning have been fighting this protracted battle." Furthermore, "after the fossil remains of the old bachelor President [Buchanan]," who vetoed the bill setting up agricultural schools in all the states, should have been forgotten, "a measure of the same character will become the law of the land."[41]

Though excusable by the poverty brought on by defeat and occupation by hostile invaders, ten years later after war had come and gone Georgia and the rest of the South had still not awakened sufficiently to have set up an agricultural college. Lee was forced to admit that in spite of all his preachments, the "Southern mind clings to traditional practices and errors with a force that nothing apparently can overcome."[42]

As the education of girls in the antebellum South could hardly be expected to extend to their attendance in a college where agriculture was taught (even if there had been one), in noting the kind of education the girls did get, Lee observed with some exaggeration, "Beating a piano with vehemence, and uttering loud and strange noises with the female voice, may be carried too far for the public good."[43] Adding another humorous touch, he inserted a squib in the *Southern Cultivator* noting that it had been proposed to establish in some Eastern city "an institution in which the science of Spinology, Weaveology, and Cookology may be taught to young ladies, and where, after obtaining these accomplishments, they may receive regular diplomas, with the honorary degrees of 'F.F.W.'—Fit For Wives."[44]

But Lee saw a hopeful sign of a better day in 1872 when a college in Illinois had set up for girls a course in which they would be able "to learn to be first-rate housekeepers—to make light-bread every time just what it ought to be—and do a thousand other things in a neat, expert and perfect way." This was a "new step in social and domestic progress," which pleased Lee, and he added, "Housekeeping as a science will be a novelty." [45] Thus were the modern schools of home economics in the making.

# ❧ III ❧

# Editing the *Southern Cultivator*
# and the *Genesee Farmer*

IN ADDITION to treating the readers of the *Southern Cultivator* to his editorials and special articles in other sections of the journal, Lee instituted a question-and-answer feature, which became a common usage in most agricultural publications. He answered these questions from his own knowledge gained by observations and experiments, from encyclopedias, from the successive agricultural volumes of the Patent Office reports, and from other sources. This practice represented an impersonal forerunner of the county agents of a century later.

Under Lee's editorship the *Southern Cultivator* improved its reputation in the South and nationally and doubled the number of its subscribers within five years—increasing them from 5,000 to 10,000 by the year 1852.[1] Adding to the excellence of the *Cultivator*, further stimulation for new subscribers came from the activities of Dennis Redmond of Augusta, who traveled over the state, not only to acquaint the people with the journal, but also to furnish accounts of any "improved mode or practice of culture, or any thing of general interest." He was soon made one of the editors, and when Lee resigned in 1859, Redmond became for a short time the sole editor and later one of two. Part of the *Cultivator*'s growth in popularity must be attributed to Lee's enlivening its pages with clear and attractive illustrations. But in November, 1851, the three-story brick building in which the journal was printed burned, destroying nearly all of the picture cuts, with the result that no illustrations appeared in the following issue. Soon Lee was able to secure a new supply, but they seemed hardly as attractive as the old ones.[2]

When Lee came South in 1847 to edit the *Southern Cultivator* and become an associate editor of the Augusta *Chronicle & Sentinel*, he did not give up his editorship and part ownership of the

*Genesee Farmer.* To a person of Lee's mental and physical vitality and verbal facility, coupled with his pleasure in traveling, there was nothing very remarkable in his editing two journals located almost a thousand miles apart. His intellectual honesty saved him from any attempt to slant his two journals to suit a Northern and a Southern audience. The agricultural programs and fundamental principles which he advocated were based on natural laws, and apart from their relationship to climate, soil, and topography, they were the same everywhere. Irrespective of North or South, water ran downhill, the sun rose in the east and set in the west, plants grew out of the ground, and chemical elements always and everywhere reacted the same way. Northerners and Southerners might differ as they pleased in politics, but in editing his farm journal Lee remained neutral.

The original *Genesee Farmer* had been started in 1831 in Rochester, New York, by Luther Tucker, an outstanding agricultural leader of his day. In 1839, he bought the *Cultivator*, published in Albany, and combined his *Farmer* with it, giving up the name of *Genesee Farmer*, and continued to publish the *Cultivator* in Albany. The next year John J. Thomas and M. B. Bateman began a journal in Rochester, adopting the name *New Genesee Farmer*, but after a half-dozen years they dropped the *New*. It now became the *Genesee Farmer*, having no connection with the first *Genesee Farmer*. Later this led to some confusion and arguments.[3]

It was this second *Genesee Farmer* with which Lee became associated. Strangely enough, he never mentioned in either the *Southern Cultivator* or the *Genesee Farmer* that he was also editing the other. The nearest Lee ever came to indicating his connection with the *Southern Cultivator* was his announcement, written from Augusta, Georgia, that since he had long wanted to study "the *art* of Husbandry" in every state in the Union he was now able to take advantage of the opportunity "of learning what he may from the experience and practice of Southern planters." Then in this letter and for some years thereafter in communications to the *Genesee Farmer*, which he published under the title of "Editorial Correspondence," he described conditions in the South and especially opportunities for Northerners to move there.[4] As he traveled around extensively, he praised the regions he visited. In one of his articles he said Southern Appalachia was a fine country, mountainous with beautiful grassy valleys, an excellent region for Northerners to settle in. He entitled this article "Wool-growing and Stock-raising in the Mountains."[5] He praised the Southern high-

lands in contrast to the sickly coastal region with its rank fertility but also with its muck and decaying matter; by "going back into the highlands you find a region perfectly healthy, where all useful grasses flourish in great luxuriance; where crystal springs and noisy brooks abound on every side," and the husbandman escaped the cold Northern winters. It was a great cattle and grape country where butter and cheese and wine could be made readily.[6] In an article entitled "Virginia Lands and Farming" he praised the opportunities that lay south of the Potomac and explained that Northerners by settling there might "do better than the best did in New York forty years ago."[7] Referring to the South as a whole, he said, "This is a good country for poor northern men, if they are only steady and industrious. Labor is not looked upon as disreputable. On the contrary, white laboring men are more esteemed here than at the North."[8]

Referring more particularly to Georgia, where he did most of his traveling, he sent dispatches back to the *Genesee Farmer* from various places in the state. He had a special liking for the uplands, the "Cherokee Country," where the Cherokees had recently lived before being sent west of the Mississippi; and after seeing this region, he could readily understand why the Indians were reluctant to go.[9] Here and elsewhere in the South he said, "A northern dairyman could make a fortune" by producing butter and cheese.[10] To his knowledge there was only one farmer in all Georgia who made a single pound of cheese, and he was from New Jersey. At this time (1847), cheese was selling at fifteen to eighteen cents a pound, and there was a ready market for it.[11] In 1848, Lee visited an agricultural fair in the town of Stone Mountain, near the famous mountain of the same name, and became almost as much interested in the mountain as in the fair. The mountain was "a great curiosity," an upthrust of granite resulting from volcanic action. He was impressed by its gigantic size, geology, plant life, and mosses and black mold "derived from the debris of cryptogamic plants."[12]

For the enlightenment of his *Genesee Farmer* readers, Lee made excursions into South Carolina to report on some of the great plantations there. In the Abbeville district, he visited the 4,000-acre plantation of George McDuffie where he learned of the system of slave labor employed there. The slaves worked by the task plan, which made it possible for a slave to finish his task by two o'clock in the afternoon, and if time off were allowed to accumulate by the week, the slave would have two days off which he could use as he pleased. Lee was much attracted by what he saw, and before returning to Georgia, he lectured in the Abbeville courthouse.[13] On another trip in South Carolina, which he reported to his Northern readers, he visited the plantation of James H. Hammond.[14]

Lee was trying to indoctrinate thoroughly his Northern readers in Southern ways of life, the crops and how they were raised, and the regional political and social philosophy. He described the annual barbecues which the great planters gave their slaves after their crops had been "laid by." There they feasted on "fat calves, pigs, lambs, kids, turkeys, geese, chickens, doves, quails and fish."[15]

Of course, Lee did not spend all of his time in the South during these years, for he was editing both the *Genesee Farmer* and the *Southern Cultivator;* of course, he did not fill his *Farmer* with dispatches from the South—he was editing a normal Northern agricultural journal and had to be in Rochester some of the time. He traveled extensively in the North, out into the Midwest and into the New England states. He visited Sandusky, Chillicothe, Cincinnati, Detroit, Maysville (Kentucky), and Boston.[16] He visited the District of Columbia, too, where he had a farm.[17] He said that he received from forty to fifty letters daily at his Rochester office, and those not taken care of by his office assistants accumulated for him to answer.[18] In addition to all of this and to his duties in Rochester and Washington, there is evidence that he entertained the plan of attending the great World's Fair in London with its famous Crystal Palace. But it turned out that Patrick Barry, the horticultural editor of the *Farmer*, made the trip and reported back what he saw there and what he saw when he then visited the Continent—all of which Lee published in his *Genesee Farmer*.[19]

In 1850, James Vick, Jr., was added to the editorial force of the *Farmer*, thus giving Lee more time to carry on his other activities in Georgia and in Washington and to travel to his heart's content. In addition to Vick was D. D. T. Moore, the proprietor who was listed also as an editor. Lee always took care of the front matter, provided his "Editorial Correspondence," and made other contributions, such as answering in a special column many of those forty or fifty letters remaining for his attention.

In 1849, this announcement appeared: "Dr. Lee has recently returned from the South, and will hereafter devote his principal time and attention to this journal."[20] It would seem that he was about ready to move back at this time and give up his Georgia connection, although there was nothing in the *Southern Cultivator* to so indicate. But there was strong evidence of the fact then in the making. In 1845, when he had become editor of the *Genesee*

*Farmer*, the financial arrangements which were made gave him "one half of the subscription right," and he promised that he would hopefully continue to toil with his "feeble pen and . . . feebler tongue."[21]

That which was in the making was the consummation of his purchase of the *Genesee Farmer* from Moore. Beginning with the January, 1850, number, Lee was listed as "Editor and Proprietor." In his announcement of the new developments, he said that he had "purchased the type, good will, patronage and subscription list," one-half of the subscription list having belonged to him from his first connection with the journal.[22] Lee was quite elated by this development. He said that his son would soon be taught to set type and run the printing department. The power press, which he now owned, would run off 2,500 sheets an hour.[23] Lee was not quite correct in his statement, for this was, in fact, a part of the purchase that went with the *Rochester Daily American* which Lee and two associates bought at this time.

When Lee first became an editor of the *Genesee Farmer*, the subscription list was between 3,000 and 4,000; since he was to own one-half of the list, it behooved him to work hard to build it up. By 1848 the number of subscribers had jumped to 20,000, which, Lee said, was the largest number of subscribers of any agricultural journal in the United States. The next year, when he became sole owner of the journal, he had a double incentive to push the list up further. After his first year as editor and proprietor, he doubled the circulation; the next year (1853) he ran it up to 50,000.[24]

The excellence of the *Farmer* helped to increase its subscribers, but even more helpful was its extremely low price of only fifty cents a year. Lee had as a fixed principle the desire to provide reading matter for farmers at a cheap price; it was ever in his mind to raise the standards of farmers and of their farming methods. What could help more than putting within their reach a good farm journal? Although the *Genesee Farmer* was now the envy of other farm journals, some attributed its large circulation more to its cheap price than to its excellence. The *Prairie Farmer* in Chicago remarked that fifty cents a year made Lee work for nothing. Lee replied, "One who is willing to work twelve hours every day . . . is not likely to beg for his bread," and he added that his *Farmer* was "the cheapest agricultural journal in the world." As to the Chicago editor suggesting that there must be "a screw loose" somewhere, Lee said that he thought the loose screw must be "somewhere out west."[25]

In 1853, when Lee's duties in Washington were coming to an

end, it seemed more than ever that he was thinking of going back to New York for good, for he wrote that he was "about to resume our researches in Western New York." But the prospects looked dim when he began to search for financial help. Heretofore he had never been able to get aid from the United States government, from the state of New York, or from any agricultural society. He had paid his expenses out of his own pocket, and some of them had been heavy, for he had had to buy expensive chemical apparatus from Europe. Woefully he observed, "Without reward, or the hope of reward, we gave months of labor, to say nothing of costly chemicals consumed, to show the relation that clover bears to wheat, and both to the soils of Western New York."[26]

Now he was appealing to one thousand men each to give one dollar a year to promote his researches in his western New York projects which he had never been able to complete. His campaign fell flat; and now in great disappointment he disconsolately wrote, "If it is not possible to raise a thousand dollars a year to promote agricultural science in the United States, the historical fact shall be recorded to the enduring disgrace of the age in which we live."[27] In 1854, he sold his *Genesee Farmer*, never to write for it again and to be ignored by its future purchasers until it came to an end in 1865.[28] It was not entirely Lee's disappointment at his failure to secure support for his researches in New York that led him to sell his *Genesee Farmer* and irrevocably to cast his lot with the South.

Lee was an indefatigable laborer both with mind and body. His pen was never still, nor were his tongue and legs, but he was not a scribbler, a gabbler, or a walker without a purpose. Although an editor had little office help in those days, editing two monthly journals and a newspaper was hardly a full-time job for him, and as he could never be still, he was destined to take on before the outbreak of the Civil War five other positions. In all he held eight, spaced so that he was to occupy for parts of the time during the period 1847–1854 four of them simultaneously; from 1854 to 1862 he would reduce the number to two.

Always bubbling over with something to say along agricultural lines, he could not confine his outpourings to the *Southern Cultivator*, the *Genesee Farmer*, and the *Rochester Daily American*, which carried farm news as well as the ordinary items which filled newspapers. Growing up in a profession which he "loved and cherished above all others,"[29] something within him seemed to force him to write as voluminously and as widely as possible—not for private gain but to help those who engaged in agriculture. As he said, "We never had any private business to promote by writing for the agricultural press, either north, south, or at the federal metropolis."[30] In seeking further outlets, he wrote in 1858 to the *Country Gentleman*, published in Albany, New York, saying that "if you will send me your weekly, you may regard me as a regular correspondent hereafter."[31] He contributed occasionally to various other agricultural journals, including the *American Agriculturist* of New York City.[32] Among his contributions to the *Transactions* of the New York State Agricultural Society was a long article entitled "The Philosophy of Tillage," which he reprinted in the *Genesee Farmer*.[33] After the Civil War he was contributor to and editor of several other agricultural journals.

The third of the four positions which he held during part of the time when he was editor of the *Southern Cultivator*, and three of them while he was still editor of the *Genesee Farmer*, was a corresponding or associate editorship of the weekly *Chronicle & Sentinel*, published in Augusta as was the *Southern Cultivator*, both being owned by the Joneses. The salary they offered which brought Lee to the South from New York was made more attractive by the addition of $2,000 which he would receive for work with this newspaper. It is not known what Lee's combined salary was; later in his career on the *Cultivator*, when he was holding some of his other positions and Redmond had been added as an editor, Lee's *Cultivator* salary was a very modest $200 a year.[34] In announcing Lee's connection with the *Chronicle & Sentinel*, the Joneses said that he would "devote the leisure that may be afforded him from the duties of that station [editor of the *Cultivator*], to writing for the columns of the *Chronicle and Sentinel*. To a cultivated mind, well versed in the principles and science of government, he adds much experience in newspaper journalism."[35] Beginning his services in October, 1847, Lee continued with the newspaper until the end of 1849. Of course, his editorial correspondence would be of the promotional kind, along agricultural, manufacturing, and commercial lines, emphasizing Georgia's resources and possibilities. In his first article he said that he would treat these subjects in a popular fashion rather than in a professional scientific account of "the dark, mysterious path in which the agriculturist is called to travel."[36]

It was only natural that he would have much to say about the present and future of Augusta, that he would praise the city as a market for planters' products and a source of their supplies, and that he would predict it would become a great manufacturing center. He advised farmers in Middle and Upper Georgia to sow more wheat than ever before, for Augusta would become even a greater milling center than it then was. He said that by 1849 its flour mills would annually need 500,000 bushels of wheat to keep running. He advised wheat raisers how best to prepare their ground for the crop: they should plow deep to "facilitate the extraction of the valuable salts of lime, potash, soda and magnesia from the sub-soil."[37] He advocated the establishment in Augusta of a "southern Agricultural and Mechanical Institute," which would be a sort of continuous fair or, and more properly, a museum where people might come and see labor-saving machinery and agricultural and horticultural products; where ladies might put on display their handiwork, such as needlework, floriculture, garden products, and canned fruits; where artists might show their paintings—in fact a complete exposition of what could be done in and around Augusta.[38]

Lee always had a sort of wanderlust in his makeup; he liked to travel. In his work on the *Chronicle & Sentinel*, he visited widely over Georgia and wrote enthusiastically of what he saw. Atlanta, although only a few years old, was a booming town, which would soon be making everything from steam locomotives to cambric needles.[39] From Marietta he wrote of the great future in store for that town and of the surrounding region's progress in livestock raising (especially in sheep), apple growing, and cheese making.[40] Lee even ventured out of Georgia as far as Detroit, where he saw a growing town with a bright future.[41]

Many of Lee's articles were unsigned, but to those who knew Lee's style of writing (and his almost inevitable comparisons with something in New York), they were easy to identify. As editor of the *Southern Cultivator* and writer for the *Chronicle & Sentinel*, he was working for the same proprietors; now and then he had articles and illustrations from the journal reprinted in the newspaper. There was logic in Lee's not signing all of his articles, for the *Chronicle & Sentinel* was an outstanding Whig and Union paper, and since Lee was a Northerner fresh from Yankeeland, it was best that some articles that he wrote be anonymous. Because Lee's province was largely agricultural, it was best that he keep out of politics in the *Chronicle & Sentinel* as he announced in the *Cultivator* that he would in its pages.

Yet sometimes Lee could not keep his Whiggish politics out of his correspondence in the newspaper. This led some of his Democratic critics to call attention to the fact that Lee was a Yankee and to claim that he had come South to engage in politics; therefore, they charged, what went into the *Southern Cultivator* might well be suspect.[42] As inactive as he was in the presidential campaign of 1848, leading to the election of Zachary Taylor, Lee received compliments for what he had done in securing the victory.[43] Looking ahead to the next election he predicted that the Whigs would win again "and thus rebuke all sectionalism and preserve the integrity and prosperity of the Union."[44]

At the end of 1849, Lee announced that he would no longer have time to write for the *Chronicle & Sentinel*, for he had received an appointment to the agricultural section of the Patent Office in Washington; but he would continue to edit the *Southern Cultivator*[45]—and for that matter, also, the *Genesee Farmer*, which he did not mention in this announcement. Lee had ten years longer to edit the *Cultivator*, and he would continue for the next four years on the *Genesee Farmer*. Furthermore, to keep his restless quill busy, during these years he would take on two other positions.[46] On Lee's leaving the *Chronicle & Sentinel* for Washington, a Georgia editor remarked, "He has no superior on the continent as an agricultural writer," and stated further that the *Cultivator* owed its high standing to Lee.[47]

# ❧ IV ❧

# Agricultural Expert
# in the Washington Patent Office

LEE'S ASSOCIATION with the Whig Augusta *Chronicle & Sentinel*, where he had displayed in his writings some of his political proclivities and had mildly aided in the election of a Whig president in 1848, and possibly his acquaintance with Millard Fillmore, the new vice president, undoubtedly had something to do with his appointment to the agricultural section of the Patent Office. Most important, however, was his reputation as an agriculturist whose passion was to help the farmer. Although in no sense had he been a political editor of the *Chronicle & Sentinel*, the proprietors of that paper were sorry to lose him both for political and agricultural reasons, and they praised his services as "most agreeable and advantageous to us, and profitable to the State."[1]

Although agriculture was the lifeblood of the nation, the government was slow to recognize or promote it. President Washington, one of the great planters of his day, seeing its importance, suggested in both his first and his last message to Congress that a board of agriculture be organized. Congress, however, did nothing about it until 1820, when the House established a committee on agriculture, and the Senate did likewise five years later.

In 1836, agriculture was given a boost when Henry L. Ellsworth was appointed Commissioner of Patents. Already possessing a wide reputation as an agricultural expert, he began collecting farm statistics and publishing in his reports articles on farming. The year following his entry into office, he started the practice of distributing free seeds. Having impressed Congress with the value of this work, he was given an appropriation of $1,000 to continue it. In 1845, Ellsworth fell a victim to politics and was succeeded by Edmund Burke, who knew little about agriculture. Solon Robinson, an able agriculturist, called Burke's first report "a bundle of trash" and the free seeds "worthless." Congress soon withdrew its appropriations for this purpose, but in 1847 it began restoring them.[2]

The condition of this department was well set forth in a biting criticism in answer to the question of where it was located: "Pent up in the cellar of the patent office, and cannot be found at midday without a candle; and when found, a single clerk struggling to get up the report."[3]

Holding Burke responsible for what went into the Patent Office reports (though a clerk in a dark cellar might be getting the agricultural information together), Lee entered the conflict in 1848 with a sharp criticism of Burke's agricultural statistics. This field had long been one of Lee's special interests, one he had tried to promote in New York before coming South. Burke had "fallen into grave errors," in his so-called statistics (which Lee called guesses) as he dealt with the production of cereal grains, their consumption, and exportation; with the subject of labor and capital; and with the value of agricultural and horticultural products. "The *guessing* at the quantity of grain grown in any State in a season," said Lee, "is so purely a matter of opinion, that it is idle to discuss the subject, with a view to remedy defects." But to be specific, in certain instances Lee said that Burke's price of wheat was a third too high and of corn a fourth. Burke set the price of hay at eight dollars a ton, but Lee said it was four dollars on an average. "Agricultural statistics," said Lee, "are a subject which has ever been shamefully neglected by American politicians." He hoped that the United States census for 1850 would collect statistics on "all the agricultural, horticultural, mechanical and manufacturing products" of the country.[4]

With the coming in of the Whigs in 1849, the Patent Office was placed in the new Department of the Interior, and Thomas Ewbank, who was born an Englishman, became the new commissioner. But apparently he was no improvement over Burke, for very soon Lee was attacking Ewbank and especially his circular which had been sent to farmers asking for information on a great many subjects. These circulars were distributed indiscriminately, helter-skelter, pell-mell, and those who received them were asked to suggest others to whom the circulars might be sent. There were twenty-four items, relating to various aspects of crops, livestock, and even to rainfall and temperature, most requiring answers which could be little less than opinions and guesses from anyone who was not an expert. Questions concerning crops probed such areas as amount of each crop raised, methods used, plant diseases, best soils and fertilizers, when they ripened, how marketed, increase or decline in amount raised, new varieties, average yield per acre, and cost of production. Questions concerning livestock covered the following areas: *cattle*, number in state, average value at three years old, cost of keep for a year; *sheep*, amount of wool clipped in a year, average weight of wool from different breeds and how marketed, number killed by dogs; *hogs*, cost of production and price of pork per pound. As for weather: how much rain fell each month, and for the same periods what were the high, mean, and low temperatures? Ewbank admitted that a person might not be able to give information on all of these items, but he should answer where he could.[5]

Lee was sorry that such a silly circular should be sent out under a Whig administration. Referring to previous agricultural material that had been published in the Patent Office reports, he said, "intelligent farmers of the country are heartily sick of these Patent Office humbugs in the shape of Agricultural Reports, written by lawyers or mechanics who are alike innocent of any knowledge of the practice of the science of Rural Economy." This work should be done by experts, and although Lee did not consider Ewbank an agricultural expert, he was quick to add that he had no "unkind feelings toward the excellent and learned Commissioner of Patents."[6] Although Lee would later change his opinion of Ewbank, now he thought that if the commissioner were not kept busy issuing patents to inventors, he should use his spare time in collecting statistics on "Mechanical Arts," with which he was familiar, "and spare the noble profession of Agriculture the infliction of dangling forever at the tail of all other interests, which are deemed worthy of attention at Washington."

Continuing, Lee said, "This course is disgraceful to a nation of intellectual farmers. It is injurious to the country, by giving to *bad guessing* a *quasi* governmental endorsement, and misleading thousands who place confidence in the wildest statements of this undigested mass of contradictory opinions. . . . If agricultural statistics and reports are worth the expense of collecting the one, and of writing and printing both, then we respectfully submit that the business should be in the hands of a gentleman who is known to the great farming interests of the Union, as one conversant with its agriculture both as a practical art and a profound science." This comment might seem to indicate that Lee thought that he himself was the man.

Referring to Ewbank's circular, Lee said that no farmer or planter "who had kept pace with the rapid progress of his profession" could read the twenty-four items on which information was wanted "and not be forcibly impressed with the conviction that the writer knew very little about the subjects treated of." As a good example he cited this one: "*Butter*. Quantity made in your state; average annual produce per cow; are cellars or spring-houses preferred?" This led Lee to remark "If public men were better informed, they would not be guilty of the supreme nonsense of asking farmers officially through the Commissioner of Patents, *to guess* how many pounds of butter are made and eaten in their respective States, in the course of a year? and what is the product of each cow?" Not one person in ten kept such records, Lee observed.[7]

A special "Bureau of Agriculture" should be set up in the Department of the Interior, Lee offered as a solution, agreeing with the *National Intelligencer* and others. Having in mind that the Commissioner of Patents was being made "jack of all trades," Lee said, "No one is expected to be skilled in several distinct trades or professions; and all that we contend for is, that the profession of agriculture shall be left to the custody of men who have devoted their lives to the diligent study and practice of this branch of human industry."[8]

The criticisms by Lee and others against a system which tried to make the Commissioner of Patents an agricultural expert reached the ears of Thomas Ewing, the Secretary of the Interior, who had the duty of setting up this new department. In a conference, Ewing and Lee came to an agreement that Lee would accept the proffered appointment to take control of the agricultural activities allotted to the Patent Office. Secretary Ewing had instructed Thomas Ewbank, the newly appointed Commissioner of Patents, to see that the work of "collecting and arranging" the agricultural material in his reports "be committed to a practical and scientific agriculturist," and very probably at this time Ewing appointed Lee to the position or suggested that Ewbank do so. At the conference it was decided that Lee should receive a salary of $2,000 annually as soon as Congress should make the necessary appropriation to the Patent Office for that purpose. Lee had insisted on that amount to replace the salary he was giving up in Georgia.[9]

On November 16, 1849, Lee announced his appointment, which, he said, had been made without application by him or by any of his friends. He was not giving up his editorship of the *Southern Cultivator*, and as this announcement was made in that journal, he did not add the fact that he was also keeping his position as editor of the *Genesee Farmer*.[10]

Lee was given no special title in the Patent Office, but he generally referred to his position as being in charge of the Agricultural Department of the Patent Office, and sometimes he was referred to by others as a clerk in that office. His appointment was welcomed generally, echoing the sentiment of a Georgia editor who wrote, "I know of no man in the Union, so competent to sift the gold from the dross, or the 'chaff from the wheat' " as was Lee.[11] As Lee put it in his announcement, he had received an "appointment to look after the agricultural matters incidentally attached to the Office of Commissioner of Patents."[12]

Lee's prescribed duties were to send out circulars to farmers all over the country asking for information, analyzing their answers, and preparing the results for publication in his reports; to distribute widely free seeds; and at his pleasure to write articles for his reports as well as to include articles written by others. Thus, in his "spare time" he edited two farm journals and gave some attention to the *Rochester Daily American*; managed a farm which he bought in the District of Columbia; traveled intermittently to Augusta and to Rochester, wandering off the direct route sometimes to observe agricultural conditions and make speeches; and engaged in activities in Washington apart from his duties in the Patent Office. As an indication of his having not left the editorship of the *Southern Cultivator*, in the March issue of 1852 he contributed six articles. Yet he did need an expert besides himself in his Washington office, and he made an effort to bring in a young agricultural chemist from England who had applied for the position and was willing to come for a salary of $1,000 a year, but Lee was unable to find the money.[13]

Beginning with Lee's advent in the Patent Office, the report of the commissioner came out in two volumes for the first time, the second volume being devoted entirely to agriculture. Among the special articles Lee contributed were: "Agriculture and Agricultural Education," "American Agricultural Literature," "Progress of Agriculture in the United States," "General View of Agriculture," "Agricultural Statistics," "Agricultural Meteorology," "Preparation and Use of Manures," "Study of Soils," "Culture of Wheat," "Culture of Indian Corn," "Potato Culture," "Cotton and Cotton Culture," and "Charcoal and Water." Some of these articles Lee republished in the *Southern Cultivator* and in the *Genesee Farmer*, and the Library of Congress listed in its general catalogue a half-dozen of Lee's articles, which apparently were reprinted as separates.[14]

The most time-consuming of Lee's work was the sending out of circulars calling for agricultural information and the processing and digesting of answers. He built the circulation list from 400 to 8,000. In 1852, he received enough answers to make two volumes each of a thousand pages. In this instance he edited, sifted, and combined them into one volume of 579 pages.[15] Ewbank, jealous of his authority as commissioner, was loath to let Lee have complete control of agricultural matters. Suppressing Lee's name, he illogically used his own at various times. For instance, he had the circulars sent out under his own name; in fact, he did not give Lee complete control over wording them. The one which Lee had so severely criticized before coming to the Patent Office was sent out again with only slight changes, which undoubtedly Lee had induced Ewbank to allow. The one sent out in 1851 was considerably amended, for the census of 1850 was then available and covered some of the items heretofore included—the one on butter was omitted and a few new items, as one on fruit culture, were added.[16]

The distribution of seeds was, of course, little more than a clerical job and required none of the special abilities which Lee possessed. This work had been started by Ellsworth back in 1836 and was then done without special appropriation, and little money was available during the time Lee had the work in hand. By this time it was beginning to take on the character of a political handout, both to promote votes for Congressmen and to curry favor with selected seedmen in providing them a market. In 1850, during the time for sending out seeds, up to April when Lee made his report, he had sent out 80,000 packages.[17]

During his term in the Patent Office, Lee sought to promote the collection of farm statistics, not only in answers to the circulars, but in every other way he could think of. He had long been advocating farm statistics. Back in 1845, he had induced the New York legislature to make provision for the collection of statistics on the various crops raised in that state; and now the campaign he had been carrying on in the Patent Office and previously had had the effect of widening the farm and land statistics in the census of 1850.[18]

In his first report Lee said, "Truthful statistics form the groundwork of all reform—of all progress," and he added that no state legislature or Congress had ever made an appropriation for the systematic collection of farm statistics, probably regarding what New York had done as not amounting to much, and, of course, not knowing what the census of the next year would include. What had been passing as statistics was nothing but crude guesses. "So far as reliable statistics are concerned, all our farm operations are conducted in midnight darkness." The number of sheep sheared had never been determined, and no count had ever been made of how many cows had been milked in any year. Nobody knew how many tons of hay or bushels of grain had been grown in any state (not excepting New York), and how many acres had been planted in cotton in the South. So, in his first report he noted that no space had been allowed "for mere *guesses* at the quantity of grain and other crops grown in the year 1849." From innocent statistical guessing, Lee said it was only a short step to falsifying statistics for argumentative and ulterior purposes; later, when Hinton R. Helper in his *Impending Crisis* had fortified his attacks on slavery with statistics, Lee put him at the head of that kind of impostor.[19] Lee would have heartily agreed with the later expression: "There are lies, damn lies, and statistics" (when cleverly manipulated).

Another subject which Lee pursued as head of the Agricultural Department of the Patent Office was war on insects. However, he was more interested in identifying insects and describing them than in anticipating the chemical warfare against them of a later time, chemist though he was. He was early and late interested in agricultural entomology, and during his Washington days he put a greater emphasis on the subject. His approach was entirely practical and utilitarian—it was applied entomology to help the farmer, not pure entomology for the sake of knowledge. In 1853, he said, "For some reason Agricultural Entomology is less studied and understood than almost any branch of rural knowledge."[20] Almost forty years later he was seeking a remedy for the buffalo gnat and calling for more study of insects and parasites,[21] noting that he had "data in hand sufficient to fill a small volume, treating of parasitic fungi."[22]

But now in 1849, Lee complained, "Insects, like wheat flies, weevils, and moths [earlier adding to his list Hessian flies, curculios, armyworms and bollworms] are annually destroying crops to the amount of millions without the least public effort being made to lessen the evil. Does it argue well for our general intelligence that no agricultural society in this most prosperous republic has ever appropriated a dollar to foster the study of Insects?"[23] "If a pirate on the high seas," he said, "or an Indian savage on land, injures the property of a citizen to the amount of a few dollars, millions are expended, if need be, to punish the offender." But what was the government or anyone else doing, he inquired, to put a stop to these insect ravages?[24]

One important remedy, he said, was the protection of birds, for most of them feed on insects and worms, thus keeping the insects from taking possession of the earth and preserving the balance that nature had set. Man could hardly "exterminate the birds of a country and not, in effect, augment indefinitely all the insects that prey upon his crops, and greatly annoy his domestic animals." Furthermore, if "it were not for the fact that insects destroy one another, and thus keep down their numbers, they might, perhaps, entirely exterminate all other living things, and then die from starvation, leaving not a plant or animal on the globe." The reproductive powers of insects were incredible, especially to anyone who knew nothing about entomology. There was "not an animal nor a plant known to science" upon which no insect subsisted.[25]

In spite of all this, Lee said, "Many boys are apparently educated to kill all the small birds that subsist on insects, so soon as these youngsters are large enough to shoulder a gun." If farmers continue to "fold their arms and say that nothing can be done, by the science of entomology, or by any other means, what but an increase of the evil is to be expected?" If no one tried, it would be "treating one's enemies with unmanly forbearance" and would evince "a belief in fatalism worthy of a disciple of Mohammed."[26] It had long been a "source of deep regret" to Lee "that the study of vegetable psychology, and of the diseases incident to cultivated plants" was generally "so little relished."[27]

Going specifically into a discussion of some insects, Lee had much to say about the Hessian fly, which, it was generally believed, had been brought ashore in straw when General William Howe invaded New York in 1776. This pest was especially bad on wheat, but Lee knew of no remedy against it, except the possibility of planting the grain early to outgrow the fly or to leave infested land free of wheat for two or three years.[28]

A calamity that was upsetting Northern farmers most in the 1840s and 1850s was a potato wilt or rot. Lee, like many others, went into the subject and said much about it; always ready to investigate as well as write, he soon discovered that a beetle punctured the stem and laid its eggs inside. At this time he was unable to find a remedy. He had made this discovery by 1845, and when a contributor to Lee's Patent Office agricultural report in 1849 sent in an article detailing what Lee had found out four years previously and setting himself up as the discoverer, Lee refused to publish the article.[29]

Lee could never remain in a place long before he acquired a

farm, for he liked to see things grow and he was an inveterate ex-
perimenter; and all of this could be best done on his own land. Soon
after becoming the commissioner of agriculture in the Patent Office, Lee bought a 110-acre farm nearby in the District of Columbia,
pretty much worn-out by tobacco culture. His son became the manager (and later probably the owner), who looked after the details
when Lee was not around; but Lee for years thereafter often visited
Washington to take a look at this farm. He was especially interested
in building up the land, and to do so, he sowed much of it in grass,
orchard grass, and timothy, but also in clover. Also he planted crops
such as corn. He said that his farm was "dedicated to experimental
purposes." During the last year of the Civil War, he was selling rye
straw in Washington at thirty dollars a ton.[30]

But Lee's private farm was not enough; the national government
should go into experimental farming and pay more attention to the
promotion of agriculture in every respect. He said, "To elevate agriculture, we must apply to it the inductive system of philosophy, and
take little or nothing for granted."[31] To accomplish this, it would
take vastly more than "a mere clerkship under a Secretary," which
Lee was intimating was his position in the Patent Office; there
should be a Board of Agriculture, as Washington had recommended, composed, Lee advocated, of "distinguished characters,"
not politicians—a board aided by government appropriations.[32] It
might be termed the "*National Academy* of Agriculture," char-
tered by Congress, and endowed by the proceeds of the sale of a
million acres of public land. It should "be to rural sciences and arts
what the West Point Academy" was "to military tactics and sciences." It should have two divisions: "Agricultural Engineering"
and "Breeding and Improvement of Live Stock."[33]

Getting down to something more practical and concrete, Lee set
his eyes on Mount Vernon as the spot which the government should
buy and develop into an experimental farm. He noted that there
was then being slowly built in the city a mighty monument to
Washington to commemorate him as a general, but nothing had
ever been done to honor him as a planter and a promoter of agriculture, nothing "but to permit the once fine estate of Mount Vernon to grow up in briars, bushes, and pines, a harbor for wild
beasts." Spending millions on the army and navy and on political
schemes and politicians' salaries, it seemed "extraordinary that a
nation of farmers" could "not afford the few dollars necessary to
make the estate of the great and good Washington an experimental

or a model farm."[34] Marshall P. Wilder of Boston, a good friend of Lee's, vigorously supported Lee in this plan of saving Mount Vernon for a good purpose.[35]

While in the Patent Office, Lee frequently saw George Washington Parke Custis, the owner of Mount Vernon, and discussed with him its possible sale to the government for an experimental farm. It seems that at about this time (1851) the estate was under contract to private investors for $200,000, with the proviso that the national government might have it at the same price, but it never acted. Lee had so long been trying to get Congress interested in promoting agriculture that he soon gave up, saying, "But it is a foolish waste of time and of energies, to go to a political Congress for any assistance whatever."[36] And so, it was left for Mount Vernon to be saved for another purpose.

# ❧ V ❧

# Activities in Washington
# and Dissensions in the Patent Office

IN THE national capital, the seat of the mighty, Lee was bubbling over with ideas. His Mount Vernon plans made little or no progress, but another dream which had long occupied his thoughts, and which he now began to push hard and to get other people interested in, was the organization of a national agricultural society or central board, which would be a federation of the three hundred state and local agricultural societies scattered throughout the nation. The separate societies, he said, "throw away their labors by isolated and independent efforts."[1] Combining units with the same interests now being the order of the day suggested to Lee that farmer organizations should do the same thing. "The science of combinations," he said, was "as applicable to agriculture as to any other business pursuit whatever."[2]

He began this preachment widely over the country: in the Nashville (Tenn.) *Banner*; the Charleston (S.C.) *Courier*; the *National Intelligencer* in Washington; in other newspapers; in farm journals; and in the Patent Office reports. Lee thought that such an organization might be called the U.S. Congress of Agriculture, and the duties of such an organization were quite evident. It could pressure Congress to make appropriations to support the work of this association. One of its duties would be to publish agricultural statistics and other information immediately, not waiting a year or two as was the custom of the government in publishing its documents. Thus, the information would be disseminated much more quickly than the Patent Office reports could be made available. "No body of men but an American Congress," said Lee, "would think of making annually a large number of almanacs, keeping them until one or two years after date, and then claiming great liberality for giving them away."[3]

Lee got his friend Marshall P. Wilder of the Massachusetts Board

41

of Agriculture to issue the call for a meeting of delegates from all states to be held in Washington. Lee insisted that it take place further south than in Philadelphia, which had been suggested, for he wanted to avoid appearances of sectionalism. To lend further importance to the meeting, presidents of agricultural societies in Pennsylvania, Maryland, and in other states signed the call. All delegates to the convention were asked to send their names to "Daniel Lee, M.D., Agricultural Department, Patent Office, Washington."[4] Later there was an attempt made to give Wilder credit for suggesting the meeting, but Lee brought out ample proof, including a statement from Wilder himself, that Lee was the originator of the meeting. In fact, four years earlier Lee had attended a meeting in New York City where delegates from fourteen states gathered for the purpose of founding a national agricultural society. It was then decided that the time was not ripe for such a move, but a committee of three was appointed, with Lee as chairman, to act when it seemed that the country was ready for such an organization. Lee had now acted, and he was especially anxious that the South be fully represented.[5]

In June 24, 1851, delegates numbering 152 from twenty-three states assembled in the Lecture Room of the Smithsonian Institution and founded the United States Agricultural Society. There were not many delegates from the South as this was considered largely a Northern movement, though Jacob Thompson, who was in Washington having recently ended a congressional career, represented Mississippi. Senator Stephen A. Douglas represented Illinois.[6]

Lee was made one of the two secretaries of this organizational meeting and he was elected Corresponding Secretary of the Society for the next year. He was, of course, prominent in the deliberations resulting in the writing of the constitution of the Society. The main purpose of the Society was to be carried forward by a board of agriculture, which was to be composed of three members from every state, territory, and from the District of Columbia. It was charged with the duty of promoting agriculture in every way: specifically through advocating agricultural schools, new crops, a geological survey, and the collection annually of agricultural statistics from every state.

There was much discussion over the subject of membership dues. As might be expected, Lee argued for low dues, for it was one of his most fixed principles that as many people as possible should be reached with agricultural information. He believed that the most

effective way to secure governmental aid was through getting farmers elected to Congress, and that could best be done by waking up
rural people through agricultural publications. "Agriculturists
love their profession," he said, "and are ready to combine their
individual efforts to elevate it, until there shall be at least as many
farmers in Congress as lawyers." He wanted the annual dues to be
one dollar and life membership ten dollars; but the dues finally
fixed were two dollars and twenty-five dollars respectively.

The visible work of the Society was to be carried on through a
quarterly journal, which after the first year became an annual publication. Lee edited the first number of the quarterly, which ran to
144 pages. In his introduction, he argued for congressional support, but despairing of such luck, he sought to set up an endowment, the income from which would free the Society from any
government help. Always a little impractical in his enthusiasm for
such projects, he wanted every state to secure 1,000 life members at
the set rate of $25. This would provide an endowment of $775,000,
which invested at 6 percent would produce $46,500 annually. This
amount would make possible the promotion of much research and
experimental activities and leave sufficient money for a quarterly
publication—agricultural news while it was fresh, not like Patent
Office reports a year or two late. Here Lee lost again, for his endowment did not succeed, and after the first year, as already stated, the
publication became an annual.[7]

Disappointed in the way the Society, which he had done so much
to found, was being run, Lee soon became lukewarm in supporting
it, and by 1854 he had ceased to be a member. Editing only the first
number of the quarterly, he was succeeded by another in the secre-
taryship after the first year. He castigated the new secretary and the
annual publication and declared that the two-dollar dues greatly
cut the prospective membership and thereby lessened the usefulness
of the Society.[8]

Although Lee was not a moving spirit in the founding of another
agricultural society, he was in complete harmony with it. This organization was the outgrowth of the feeling that the United States
Agricultural Society was a Northern affair and was run by Northerners; and, as Lee believed, it was beginning to propagandize
against the South and its slave agriculture. The new society was
founded in Montgomery, Alabama, in 1853 under the name the
Agricultural Association of the Slave-Holding States. This title was
soon changed to the Agricultural Association of the Planting States,
but as its first name stated, it was to be confined to the slaveholding

states. In answer to the charge that "the world is against us" the Executive Council said, "Be it so; the world, we know, is dependent on us, and we glory in our position. Let us be true to ourselves, and all will be well." The purpose of the association was to diffuse knowledge in all branches of agriculture and to defend "our peculiar institutions." It desired a close bond of fellowship to develop the South's resources "and be united as one man in our interests." George R. Gilmer, a former governor of Georgia, was the president. Its first meeting was held in Macon, Georgia, and the next two took place in Montgomery and in Columbia, South Carolina.[9]

Lee was always proud of his work in the Patent Office, of the policies he pursued, and of the programs he advocated. It became almost traditional to say that his advocacy of agricultural schools led to the appropriation of public lands for that purpose in the Morrill Act of 1862.[10] His first volume (Volume II of the Patent Office report for 1849) contained 574 pages with seven plates of illustrations. Similar volumes followed for the next three years; and all told, Congress printed and sent out 540,000 copies of Lee's four volumes. This wide distribution greatly pleased him. Some of his articles were reprinted in farm journals and newspapers and were quoted and favorably mentioned in England.[11]

Lee's leaving the Patent Office did not come under pleasant circumstances. It seems that his salary was not a special appropriation for the Agricultural Department but was allotted from the total appropriation for the Patent Office, which was under the control of Commissioner Ewbank. Being jealous of Lee and never quite approving of him, Ewbank was slow to allow him the $2,000 annual salary which Lee said Secretary Ewing had promised him. Lee charged that Ewbank had deprived him "of nearly a whole year's salary." He believed the reason Ewbank had done so was because Lee had shown Southern agricultural production, under the slave system, in a better light than had been done in previous reports.[12]

Lee was forced out before the completion of his fourth year, as he later charged, "to make room for one who was willing to prepare anti-slavery statistics." He held that Patent Office agricultural statistics after his departure had been manipulated as antislavery propaganda to show that slave labor reduced the yield in agricultural products over that of free labor.[13] He said that some of these statistics were so crude as to be contradictory and unbelievable to intelligent people.[14] Entering into the picture was William S. King, the editor of the *Boston Journal of Agriculture,* who was publishing "false and calumnious accusations" against Lee to so discredit him that King might get the position.[15]

Lee might well have expected to go out with the expiration of the Whig administration of President Fillmore and the entry of the Democrats under Franklin L. Pierce, for politics permeated government service. He vigorously denied the charges of the *Albany Evening Journal*, the *Rochester Democrat*, the *New-York Daily Tribune*, and other Northern newspapers, as well as a few Southern ones, that he was a partisan of the late Whig administration and "a particular favorite of Mr. FILLMORE." King in his agricultural journal charged Lee with being "Our Caleb Quotem, who is at once an M.D., editor of two agricultural papers, publisher of one or two political sheets, and *political stump-speaker*, Agricultural Clerk in the Patent Office, and seed-raiser in general to the same."

In answer to these charges, Lee said that "this 'M.D.' is an industrious man, which is more than can be said of some of his assailants." He was especially desirous of answering King's implied charge that Lee was engaged in raising seeds to sell to the government and then distributing them free to farmers from the Patent Office. King probably got this idea from the fact that Senator Pierre Soulé had introduced in the Senate a bill to make an appropriation "to produce such improved seeds of grass, grain, and vegetables, as may be wanted at the Patent Office; the same to be expended under the direction of DANIEL LEE." Soulé in this action was responding to Lee's frequent accusations that the government was buying bushels of worthless seeds and being made the victim of dishonest seedmen by having these "common and valueless seeds" distributed to the public. Lee had tried to put a stop to these dishonesties but had always been overruled by Commissioner Ewbank. Lee, of course, was raising no seeds for the government.

The controversy now extended to Ewbank who as Commissioner of the Patent Office had few supporters. Lee introduced a letter from Charles S. Stansbury, who had been associated with the Patent Office for eight years, was secretary of the National Institute, and who had been the official United States representative to the World's Fair in London. Stansbury said that Ewbank had been guilty of low practices in office, that he had made promises which he did not fulfill and then denied that he had made such promises. Stansbury would not "believe him under oath, in any case in which his passions, or his interests were concerned." Ewbank had been dismissed "when [his] management threatened the utter and irretrievable ruin of the Bureau over which he presided. In quitting it," Stansbury said, "I have yet to learn that he has left behind a friend, or carried away the respect of one of his official associates." [16]

Only after Lee had left the Patent Office did the full story of his struggle against the dictation and hostility of Ewbank come out. Had Lee known the domineering attitude of Ewbank toward all under his authority, he doubtless would not have accepted the position. It seems that no fixed term for Lee's services had been set, for he was appointed especially to get out the agricultural report for 1849; but it was his assumption that he would remain during the rest of the four years of the Whig administration. His appointment came during Taylor's first year in the presidency and directly through the solicitation of Thomas Ewing, the Secretary of the Interior; but on July 9, 1850, Taylor died, and was succeeded by Millard Fillmore, a friend of Lee's when both were living in Buffalo.

As was the custom, a new president was given the opportunity to appoint his own cabinet; and so, after an interim of a few months, President Fillmore appointed as his Secretary of the Interior Alexander H. H. Stuart to succeed Ewing. Although Ewbank was not displaced, he was not considered secure in his position for some time; and assuming that Lee had fulfilled the purpose of his original appointment, and coming to dislike Lee, he now hoped to be rid of him. Secretary Stuart in a conference with Fillmore brought up the matter of Lee's continuance in the Agricultural Department of the Patent Office, and, as might well have been expected, the President said he wanted Lee retained. Stuart communicated this information to Ewbank on October 23, 1850.[17]

Commissioner Ewbank, being greatly displeased at having to keep Lee, immediately worked up a series of charges against him and five days later wrote them out in a letter to Stuart. His overall complaint was that Lee was guilty of independency, insubordination, and refusal to obey orders. Specifically, Lee signed reports before submitting them and also signed his name where Ewbank's should have been used; he put books in the department library and paid himself for them; he had printed in his Rochester printing plant labels used for sending out free seeds; and, in general, he did not perform the duties of his "clerkship." Ewbank said that Lee could have written his agricultural reports at home (intimating that he need not have come to Washington to do so) as well as in the Patent Office; and furthermore, his appointment was only temporary, and his predecessors had performed their work within the time needed and had departed having been paid only $700 or $800 for their services. Also, Ewbank seemed to have been upset by Lee's frequently advocating that Congress set up an agricultural bureau, fearing that this would deprive him of his authority over agricultural matters in the Patent Office.[18]

Ewbank's letter did not lead to Lee's dismissal, but it did cause Fillmore to instruct Stuart to inform Ewbank that Lee should be retained, and it led Lee to make his devastating defense.[19] As editor of the *Southern Cultivator* and as former assistant editor of the Augusta *Weekly Chronicle & Sentinel*, he had excellent vehicles in which to carry his reply to the people, using them in the February and April 9 issues, respectively.[20] He won a point immediately with his Southern readers when he said that when he first went to Washington he lived at the same boarding house with Ewbank, and, thus, became intimately acquainted with him, his character and beliefs. He found him to be a confirmed abolitionist and atheist: "I was less surprised at the intensity of his abolitionism, than shocked at the baldness of his atheism."[21]

Lee charged Ewbank with making it almost impossible for him to secure seeds and distribute them, Ewbank generally explaining that there was no more money left in the agricultural fund—either for seeds or for Lee's salary. Thus argued this man Ewbank, "who by one of those mishaps that sometimes occurs, is a warm abolitionist and the liberal patron of his sect." Lee added that he, himself, was "not trained in that company."[22] Lee had no influence whatever with him "either in procuring or distributing seeds," although in name he had charge of the agricultural section of the Patent Office. Ewbank was "an unworthy character," and Lee had tried to have the agricultural funds removed from his control, but he had not succeeded. Ewbank, then, on the grounds that there was no more money left in the appropriation for Lee's salary, told him that he was no longer needed and was, therefore, dismissed—a move which did not succeed. (And later when Ewbank himself had been dismissed, it was charged by his friends, the few whom he had, that he lost his position because of his attempt to dispose of Lee.) But before the matter was resolved, Ewbank had tried to appoint in Lee's place a young man "who had made himself useful by franking MR. [William H.] SEWARD's 'Higher Law' Speeches in the Patent Office."[23]

But as to the distribution of the seeds, Ewbank sent them to Rochester to be packaged and be sent out in order "to bestow a fat job on a patron of FRED. DOUGLASS, and his abolitionist 'North Star.'"[24]

Lee charged that this was at a cost of twice what the work would have cost in the Patent Office. There was also the possibility that a seedhouse could substitute faulty seeds for the good ones it had received.

In defending himself and attacking Ewbank, Lee suggested that the Commissioner should use his time in processing patents and not in attempting to interfere with agricultural matters about which he knew nothing. He pointed out that Ewbank was a "city-bred mechanic" and was "a humbug of the first water in Mechanical Science." If Ewbank would devote his time to what little he knew about mechanical matters, then a system could be organized in the Patent Office free from "confusion and contemptible favoritism."[25]

Lee charged that Ewbank had not only mismanaged the agricultural funds but also had misappropriated some of the other funds of the Patent Office. In promoting his own position, Ewbank had bought publicity in certain newspapers to the amount of $231.25— of which the *National Intelligencer* received $87.00 for a defense of himself which Ewbank wrote, and of which Horace Greeley in his *New-York Daily Tribune* received $30.00 for news stories favorable to Ewbank. He had also spent another sum of $1,000 for which he could not provide a proper accounting.[26]

When Lee had sought to aid Southern agriculture by the importation of some special varieties of seeds, he was prevented from doing so by Ewbank. Lee concluded that Southern agriculture had no representative in Washington "for we have no authority to speak or act in its behalf."[27]

In fine, Ewbank was a tyrant who had the twenty or thirty clerks in the Patent Office subdued and afraid to speak up: By "his maladministration, insolent bearing and tyranny over all subordinates in the Patent Office, he has kept it in a state of civil war." He had placed on their necks "the yoke of John Bull despotism."[28]

In explaining and defending himself in his articles in the *Southern Cultivator* and in the Augusta *Chronicle & Sentinel*, Lee said that when Ewbank resolved to make the agricultural appropriation to the Patent Office "an abolition fund, we had no other choice left but to be a silent party to, or an open opponent of the scheme. We chose the latter course."[29]

Ewbank, who had begun the fight against Lee in his letter to Secretary Stuart of October 28, 1850, was greatly angered by Lee's replies in the *Southern Cultivator* and the Augusta *Chronicle & Sentinel*. Ewbank wrote Lee on April 29 to tell him that he had laid the articles before President Fillmore and Secretary Stuart, and he demanded that Lee substantiate his charges. Lee replied the same day, saying he would prove his charges when Ewbank proved his.[30]

With no love or respect for each other, Lee and Ewbank were able to get along with their work in the same building; but when both were no longer in the Patent Office, the fight broke out furiously into the open. The *Rochester Democrat*, one of the several papers of the city which had long feuded with Lee personally and disliked his Whiggish-American politics, entered the fray in late 1852. Referring to his acceptance of the position in the Patent Office, it called Lee an "office-beggar."[31] Another of the Rochester newspapers, the *Daily American*, of which Lee was part owner and one of the editors, of course, came to Lee's defense.

But to throw the fat into the fire with a vengeance, Horace Greeley took a hand. He had all along been friendly with Ewbank, who had lived in New York City where Greeley edited his *Tribune*. In three articles in his paper during December, 1852, Greeley attacked Lee, echoing much of what Ewbank had been saying.[32] Remembering Lee's course in the New York legislature in 1844–1845, Greeley called him "the wildest and most sweeping Radical ever sent there."[33] He was "a reptile"[34] and "a greedy and malicious subordinate" of Ewbank's.[35] He had been hired temporarily to complete the agricultural section of the Patent Office report for 1849 and had wormed himself into a continuing position at $2,000 a year, being only a subordinate clerk. Lee at the same time held other positions, and, even so, Greeley charged that Lee could do within four months all the work that was necessary to be done in Washington.[36]

Lee himself wielded a sharp rapier. He demanded of Greeley space in the *Tribune* to answer these attacks, and when he was denied the opportunity, he declared that Greeley was "so much of a coward as well as a calumniator, that he dare not allow me a fair hearing before those whom he and MR. EWBANK have abused by seven columns of falsehood."[37] Greeley answered by saying that he "had never seen" a letter from Lee demanding space in the *Tribune*.[38] Coming to the rescue, Alexander Mann, the senior editor of the *Rochester Daily American*, said, "Never was a baser attack made by any unscrupulous press [on Lee], to shield a fallen official [Ewbank]."[39]

Lee then proceeded in seven articles in the *Daily American* to defend himself against Greeley and Ewbank.[40] He wrote numerous people who had been under Ewbank in the Patent Office and re-

ceived devastating criticisms of Ewbank in reply, which Lee published. He proceeded to confront Greeley with the fact that he had
urged Lee to accept the position in the Patent Office, Greeley being
somewhat of an agriculturist himself and at that time recognizing
Lee as an outstanding agricultural expert.[41] As for his salary, Lee
said that there "never was a month in which I would have resigned
should I have obtained a fair settlement with the dishonest man at
the head of the Patent Office."[42] For the last seven months of 1852,
Lee had not been paid a single dollar from the public treasury, he
said, and he chided Greeley: "Why then should Horace Greeley so
wantonly and maliciously assail [me]."[43] Although Lee had been
promised a salary of $2,000 a year, for more than three years he had
been able to collect only $4,500.[44]

Although the other offices in the Patent Office had carpeted floors,
the Agricultural Department, which was in the basement, "had
nothing but a brick and naked cement and cellar bottom." Lee had
a mattress put on the floor and was forced to pay for it when Ewbank refused to let payment come from the agricultural funds. The
amount, being only $9.88, was as petty as was Ewbank's refusal.[45]

On June 30, 1852, Ewbank wrote Lee that since the agricultural
fund had been exhausted his services ended with that day.[46] Not
long thereafter, Ewbank's services were ended by the appointment
of Silas Hodges of Vermont to head the Patent Office.[47] Greeley
charged that Lee, who was a close friend of President Fillmore's,
was responsible for Ewbank's dismissal, but Lee denied having had
anything to do with it.[48] With his other jobs and no salary for the
latter part of Fillmore's administration, Lee soon faded out of the
picture.

Despite the contributions Lee had made while in the Patent Office, of which he was justly proud, he was now glad to be out of the
turmoil that had gone with it; but he would still remember Washington with pleasure and would occasionally return to visit old
acquaintances and his son, who was running the nearby farm. During his years in the Patent Office, Lee had besides his residence in
Washington two other places, Augusta, Georgia, and Rochester,
New York, and he had been visiting among them as duties dictated.
In giving up Washington in 1853, it seemed to have entered his
mind for a time that he might go back to Rochester to settle down,
for he said in his *Genesee Farmer* in announcing his departure from
the Patent Office that "we are now happy to return, after an absence
of nearly six years [reckoning from the time when he went to Augusta, in 1847], to our Western New York, to labor while life shall

last for the advancement of Agricultural and Industrial Education."[49] But this was not to be, for the next year he divested himself of the ownership of his *Genesee Farmer* and went back South for good.

# ❧ VI ❧

## Newspaper Editor—
### *Rochester Daily American*

LESS THAN two months after Lee had begun his work in the Patent Office, while he was still editor of the *Southern Cultivator* and of the *Genesee Farmer,* he became part owner and an editor of the *Rochester Daily American.* This paper had been founded in 1844, after the defeat of Henry Clay for the presidency, to advocate Whig and Native American policies. In the course of time the brothers Leonard and Lawrence Jerome became owners and sold it together with a large printing plant to Lee, Alexander Mann, and John Morey. The paper first appeared under its new owners on January 1, 1850. Leonard Jerome soon after the sale moved to New York City, made fortunes and lost them on the stock market, and added to his fame in history by becoming the grandfather of Sir Winston Churchill through his daughter Jennie's having married Lord Randolph Churchill—they met on the Isle of Wight one day and two days later became engaged.

The newspaper and the printing establishment, now operating under the firm name of Lee, Mann and Company, represented a purchase of considerable proportions. The plant had for some years been publishing not only the *American* but also printing the *Genesee Farmer* and several other journals. The editors as announced on the masthead were Daniel Lee and Alexander Mann. Morey was in charge of the printing plant and its businesses. Mann had been editor of the *American* from its beginning and was considered the senior editor, and much of the success and acclaim of the paper was due to him. As for Lee, the Jeromes said that the paper had "been enriched by his pen—and as an editor, author and legislator he is most favorably known to the citizens of Western New York," having rendered editorial service "to the great industrial interests of this section of the country." He had been an associate editor of the paper for the past years. The new owners promised to make no change in its political principles and to advocate more than ever the progress of "Agriculture, Mechanical labor, and industrial pursuits generally."[1] In the early part of 1851, Chester P. Dewey became an associate editor of the *American,* and two years later he acquired a third ownership, buying Morey's interest; Morey continued as manager of the printing establishment.[2] The editorial board was now announced as "A. Mann, D. Lee, and C. P. Dewey, Editors."

The *American* was published in three editions, daily, triweekly, and weekly at an annual rate respectively of $8.00, $4.00, and $1.50. The printing plant was extensive and supplied with the most modern equipment. It had five power presses run by a ten-horsepower steam engine, three hand presses, and other necessary machinery. Besides printing the *American,* the *Genesee Farmer,* and other journals, under the name of the American Job and Book Office it published books and did commercial work.[3]

It is evident, then, that even with his three other positions, Lee was not weighted down by editorial duties on the *American* for Mann, as senior editor, had principal charge of the paper, and after the first year Dewey was an efficient associate editor. Lee remained for the most part in the Patent Office in Washington and sent his editorials from there. It had been announced that Lee would be in charge of editorializing on agricultural and industrial subjects; but since the *American* was primarily a political newspaper and since the *Genesee Farmer* came off the same presses, the paper would devote its principal attention to politics and general news, local, state, and especially national.

Unless something of special and personal importance came up, all editorials were unsigned, but it was easy for those who knew Lee's interests and style of writing to determine which editorials he wrote. For instance, the lead editorial in the August 20, 1851, issue, "Agricultural Products in 1851," could easily be attributed to Lee.[4] As western New York was a great wheat country, especially Monroe County, Lee had much to say about that crop, its progress, its marketing, and its further promotion. In January, 1854, Lee in an editorial on wheat stated that its price was then two dollars a bushel, and he praised it as a great factor in the economy of the state. He did not neglect promoting the corn crop, grape culture, sheep raising, and wool production.[5] He was loud in his praise of the agricultural wealth and possibilities of western New York: "All things considered, there is no other section so favorable to agricultural operations as Western New York. Soil, climate, proximity to market, density of population, canals and railroads combine to make it the garden of the Union."[6] Since the *American* had no subscribers in the South, Lee did not slant any of his agricultural editorials toward that region; he was taking care of their interests in his monthly *Southern Cultivator.*

As an editor of the *American* Lee spoke in two voices. His major voice sounded off in leading editorials, which were not numerous, but his most frequently heard voice came in what was termed "Editorial Correspondence of the Daily American," carrying a Washington dateline. As he was doing editorial work in publications located in three widely separated places (Augusta, Washington, and Rochester), he could hardly be long in any of the three places except in Washington, where his government position placed him. Beginning with his part ownership of the *American*, in January, 1850, he sent very frequently (almost daily for some weeks) his dispatches to Rochester, until the end of September of the same year. This period of heavy correspondence was due to the excitement in the national capital, as well as all over the country, in the session of Congress beginning in January and ending in September. This was the national crisis that was finally settled (for the time) in what came to be called the Compromise of 1850. After 1850, Lee's contributions to the *American* were much less frequent, and they practically played out in the latter part of 1854, when he accepted a professorship in the University of Georgia, Athens.

In his editorial correspondence Lee touched on many subjects, but during the compromise crisis he devoted his major attention to political developments. Nevertheless, it was only natural that he would have much to say about agriculture, since he would be filled with such information garnered from his Patent Office work. He wrote about the value of agricultural statistics; agricultural conditions, North, South, and abroad; and programs for agricultural betterment.[7] He took occasion to advocate more agricultural activities in the wide open District of Columbia, where he could not resist buying a farm and engaging in experimental planting on it. "Washington is a great place," he wrote, "and abounds in all sorts of wisdom but that which relates to the commercial production of eggs, potatoes, milk, and butter."[8]

The streets of the national capital were not clean enough to meet Lee's approval, as, indeed, they were not in most American cities; and he thought this pollution was not only a national disgrace but an ever present danger of disease and epidemics.[9] He favored increasing trade with Canada,[10] and he commented, agreeing or disagreeing, on the principal issues of the day. As many of the great political leaders finally passed off the stage, he shed tears or possibly was not too sorry that some of them were gone. He greatly mourned the death of President Taylor in early July, 1850,[11] but found some balm in that his friend Millard Fillmore succeeded to the presidency. With the death of John C. Calhoun, he was not so upset, saying that his going was "a severe blow to disunionists of the South," adding that they were "as badly off as a swarm of bees without a queen" and had no prospect of getting another leader.[12] As for the national economy, he was strongly opposed to congressional extravagance as well as too much spending throughout the country.[13]

But the perils of the Union were much on Lee's mind and conscience (not that he had helped to bring them on). He looked on Seward's "Higher Law" doctrine as next to treason, for it imperiled the Union, insulted the South, and was wholly at variance with the Constitution.[14] Lee could well see the position of the Southerners in their struggle for the protection of the Constitution, and it was time that Northern leaders quit baiting the South.[15] He warned that "without more confidence in the sense of justice, the people of the South will soon set up an independent government of their own." [16] So intense was Lee's interest in national politics and in the preservation of the Union that at times, it seems, he almost forgot his first love—the farmers and laboring people and their problems.

During the congressional session of 1850, Lee could almost be termed the political correspondent of the *American*. His paper was Whig, Lee was a Whig, and he could not forget it. A Whig administration had appointed him to the Patent Office, and the political atmosphere that pervaded Washington began to seep into him while he was a resident of the national capital. A great Whig, Henry Clay, was largely responsible for formulating the terms of the Compromise of 1850, and great Whigs like Clay and Webster (whose speeches he praised) fought for months to put it through. Although President Taylor had not favored all its terms, his death in the midst of the struggle led to the succession of Fillmore, who lent his efforts in its behalf.[17] The months of debate Lee reported faithfully, adding his own agreements and disagreements.[18]

The perils of the times, which the compromise attempted to settle, were shot through with the slavery agitation. An important element related to the extension of the "peculiar institution" into the territories that had been secured at the end of the Mexican War. Looking at the semidesert region of what would be divided

later into Utah, New Mexico, and Arizona, where nobody ever
would want to take his slaves, Lee wisely remarked that "the South
has nothing to gain; the North nothing to lose. The question is
bootless, except as it may serve to make political capital, by playing
on the sectional feelings and prejudices of the people, in either
quarter of the Union."[19]

The question of slavery loomed big in Lee's mind, and his views
brought about much disputation and personal attacks on him in-
creasingly bitter until the argument was finally settled in civil war.
He had much to say on the subject of slavery as it permeated the
debates in Congress on the compromise.[20] The general policy of the
*American* on slavery was somewhat at variance with the attitude
Lee was thought to have, especially after he became permanently
domiciled in the South in 1854. In fact, one of the editorials which
he might not have written (but which as an editor and owner of
the paper he did not veto) plainly said, "We make no apology for
Slavery or the Fugitive Slave Law [one of the provisions in the
compromise]. Human bondage we abhor. . . ."; but the law and
the Constitution should be upheld.[21] And again, in the language
of an editorial, "We deplore the existence of slavery," but it had
the protection of the Constitution, which must be obeyed.[22]

At this time Lee supported emancipation, but he opposed having
it thrust on the South by Northern abolitionists—let the movement
develop in the South, he argued.[23] He supported the work of the
Colonization Society and agreed that free Negroes should be sent
to Liberia or colonized in some other part of the world.[24] He said
that if all the money that England and New England had made out
of supplying the South with slaves had been put at 6 percent interest, the amount of money that would have resulted would exceed
"the present value of all the slaves in the United States." This
would provide the funds for buying all the slaves and sending them
back to Africa.[25] During the early 1850s he opposed reopening the
foreign slave trade, except under certain conditions; but his final
position was all-out importation of African slaves.[26]

As editor and editorial correspondent of the *American* Lee
greatly enraged the *Rochester Democrat* and, to a lesser extent,
the *Rochester Advertiser* and the *Rochester Union*, with his subjects of slavery and politics. But getting away from the turmoil of
Rochester newspaper attacks was not his reason for accepting a professorship in the University of Georgia or for selling the *Genesee
Farmer* in 1854 and his interest in the *American* in 1856. As has
been noted, he ceased to perform any editorial work on the news-

paper after 1854, and it was a matter of time until the partnership would be dissolved. The news broke on May 21, 1856, when the *American* appeared under the editorship and proprietorship of Chester P. Dewey, who had been one of the editors and part owners for the past few years. Lee was now busily teaching agriculture in the University of Georgia, and Mann moved down to Albany to edit the *Albany Statesman* and support the American Party.[27]

# ✣ VII ✣

# Soils Exhaustion and Restoration

THE MAIN outlets in antebellum times for Lee's pronounce-ments and agricultural programs were the *Genesee Farmer*, 1845–1854; the *Southern Cultivator*, 1847–1859; the Augusta *Weekly Chronicle & Sentinel*, 1847–1849; the Patent Office reports, 1849–1853; and the *Southern Field and Fireside*, 1859–1862; and to a lesser extent, the *Rochester American* (daily and weekly) with which he was associated from 1850 to 1856.

Lee had a "message" for the farmers, which he never throughout a long life tired in delivering. This is not to say that it was all new, for many of the agricultural writers were saying much the same, but Lee had certain specialties, which he might well say were his own, in that he hammered on them in season and out. He was scientific throughout; he took nothing for granted. Study and experimenta-tion meant everything; and only knowledge so gained was what should be relied upon and put into practice. Although Lee was a practical "dirt" farmer, in his writings in agricultural journals he was likely too often to go over the heads of his readers, whom he always liked to consider ordinary hard-working men of the soil.

"Both pestilence and famine," he said, were "the offspring of ignorance." "Rural science" was "not a mere plaything for the amusement of grownup children."[1] Late in life, he said that for forty years he had proclaimed the fact "(perhaps the discovery) that agriculture is a natural science because all its industrial results and processes are governed by natural laws."[2] He explained that natural laws were "the laws of Omnipotence—always right, always instruc-tive, always deserving of profound study."[3] And when he was nearly ninety years old he proclaimed the article of faith that there was "no more pleasant study to carry ones life forward to ninety years than the study of agricultural anatomy and physiology, based on chemistry, geology and climatology."[4] He would go forward along these lines, and others, too; but his first love was chemistry as ap-plied to agriculture. Although a doctor of medicine, he gave up the practice after a few years, and he never made a study of human anatomy; but his inquisitiveness led him to speculate on the causes of some diseases. He investigated cholera sufficiently to say that it was caused by conditions such as filth and the carelessness of people in disposing of waste and trash. He declared that good health was "above all price in value."[5]

In 1852, he said that there were more agricultural journals published in the United States "than in all the world beside," and that within fifty years there would be 100 million people, and that three-fourths of them would be engaged in agriculture and horticulture —overstating the population by a dozen years too soon and not realizing the growth of cities, which would greatly reduce the rural population. But in the light of his predictions, research was vitally necessary and its lack was appalling. The repetition of known facts in all the journals in the country was not enough. "We may all repeat what little we know a million times each," he said, "and leave the sum total of knowledge just as we found it. Progress implies an advancement from things known to things unknown—an addition to the aggregate wisdom of the world."[6]

In running a series of articles on agricultural chemistry, he said that in "its application to tillage and husbandry, chemistry is so vast a subject we hardly know how to give it a proper introduction to the reader."[7] He explained that agricultural chemistry was the "art of transforming soil into bread,"[8] and he was to repeat this phrase almost endlessly—leading the editor of the *Prairie Farmer* to remark that Lee continued to write "of making mud and turf into bones and tallow and so on."[9] But another editor remarked that Lee's name was "the synonym of science."[10] Lee was not slow to admit his long involvement in the subject, describing himself in 1858 as having "probably studied the chemical properties of soils, their sources of fertility, by laboratory researches and otherwise, as thoroughly for the last fifteen or twenty years, as any other man in the country."[11] In defense of his longtime study, he said, "Without the assistance of this most important science, we should know next to nothing of the elements that form our soil, our grains, cotton, root, fruit, meat and other provisions."[12] In trying to develop a sense of curiosity, he remarked how wonderful it was that millions of people from childhood to old age ate apples but the fact could not to be bragged about since those same people never tried "to learn how an apple grows, nor to understand the nature and properties of things that really make an apple."[13] When he became the editor of the *Southern Cultivator,* he added a department which he called "Agricultural Chemistry," heading it with an appropriate illustration.

Lee gave "an all-wise and beneficent Providence"[14] credit for establishing the laws, forms, and elements in the chemistry of soils, the discovery and utilization of which were the work of the chemist. In discussing this subject under the heading of "Agricultural Physiology," he said it could not be understood "without assuming some knowledge of geology, chemistry, mineralogy and meteorology," as well as of the forms and anatomy of plants and animals.[15] Getting down to specific facts, he said that there were "some fourteen different substances which enter into the composition of plants and which must exist in an available form in the soil or atmosphere surrounding it, before any crop can be obtained."[16] The half-dozen or more "most valuable elements of all crops" were ammonia, phosphorus, sulphur, potash, chlorine, lime, magnesia, and nitrogen.[17] It took "many *things* to form a single blade of grass . . . [and] these things must exist somewhere . . . more than one half of the ash of [Irish] potatoes is pure potash."[18] About half of the dry weight of all plants, he said, was carbon or charcoal.[19]

Now, where did all these elements come from? They came from "earth, air and water" and were transformed "into cotton and gold,"[20] an answer Lee gave frequently. And, of course, he varied the expression by including other crops, and sometimes he omitted

Agricultural Chemistry.

gold as a possible final product of all crops, because he abhorred moneymaking as an end in itself.

At one time or another Lee emphasized air or water or earth as a source of plant food to the exclusion of the others as his enthusiasm for that particular source carried him forward. Air was "the great store-house of plant food."[21] At another time, he said, "Exclusive of a small portion of earthy matter, all plants and all animals are nothing more than CONSOLIDATED AIR."[22] When plants were burnt or when they rotted, most of their fertilizing qualities went into the air to be drunk up or absorbed. Lee said, "The air gives back to growing plants all that it receives from them when they decompose. Hence, the atmosphere is nature's grand store house for feeding all plants and clothing the earth with rich vegetation."[23]

Lee could be equally eloquent in praising water as a source of plant food. "Rain and snow-water, dews, and the hydrometric condition of the air, have an important and controlling influence on the growth of all vegetables," he said.[24] Much of the soluble chemicals went into the rivers and were carried to the ocean, which became the great storehouse of such chemicals as magnesia. Lee or no one else in his times, knew that in the twentieth century vast supplies of magnesia would be recovered from ocean water. "We want fields that will be perpetual fountains of milk, rich in curds and cream," he said, "and in the flesh and bones of animals sold off the farm; and we may obtain their earthly elements either from rocks and their debris directly, or from the ocean after they are carried thither by moving water."[25]

Lee liked to analyze the various waters and discuss his findings—springwater, swamp water, creek water, river water, and rainwater.[26] He disagreed with some of the early writings of Baron Justus von Liebig as to what was soluble in water and what was not, contending that the famous German chemist was wrong when he said that water could not dissolve potash, ammonia, silica, or phosphoric acid.[27] Just as Lee had praised air as the great storehouse of plant food, now, in discussing water, he said that it was "the cheapest and most reliable fertilizer known."[28] Therefore, farmers should welcome the overflow of creeks and rivers on their lands, for, unless too voluminous and swift, it deposited valuable fertilizers.[29]

These overflows were nature's way of helping the farmer, as was the chemically laden rainwater; but nature could not be depended upon to properly time these overflows as they were needed. Hence,

man could come to his own rescue by irrigating his lands, and Lee
frequently argued for this. He was not specific enough as to methods, thus provoking "Dry as Dust" from Alabama to make this in-
quiry: "What does Dr. Lee mean by his allusion to irrigation in
this country? I had, in the simplicity of not knowing, supposed that
irrigation to be practicable only in Egypt, or some other region not
favored with rain, until I read Dr. Lee and some body else about
irrigation in France, and irrigation *to be* in the United States.
Messrs. Editors, is this book farming? Dr. Lee will please tell us
how it is done. This request is made in good faith. I admire the
Doctor's spunk in the good cause, and read all that he writes that
I can lay my hands on."

In answer, Lee said that three cubic feet of water "(more or less)"
annually fell on every square foot of plantation land generally over
the South. This water could be saved in reservoirs on higher land
and then as needed conveyed in canals and ditches and distributed
on lower land in small rills through sluice gates. It was best, he
said, to open the gates in cloudy weather or at night to prevent
evaporation and the baking of the soil. "Like the goodness of God,"
running water was "an ocean of benevolence."[30]

But there were other ways to collect water to be spread over the
land. There were methods of lifting water mechanically from
streams. "Hydraulics, hydrostatics, and agricultural engineering
generally," Lee said, "should be carefully studied by every gentleman who aims to be a scientific farmer."[31] And he would not let
farmers forget: "With skillful engineering, the force that is ade-
quate to carry a soil away from a field to a distant stream, can carry
as much back again and make ample restitution."[32] An overshot
waterwheel, which provided power for grinding grain, could also
provide power for lifting water from streams, or water from the
race which ran the wheel could be diverted at that height to irrigate land. Hydraulic rams, which were coming into vogue, could
provide a small amount of water to be used for small gardens or to
be stored in tanks to be used over wider fields. Then, there was the
Cornish engine on the market, which the builders guaranteed
would lift 50 million pounds of water one foot high with 100
pounds of coal—with variations, of course, according to the amounts
of water, height, and coal. And with these mechanical lifts, Lee
would have farmers remember that the power which they developed could be used to drain wet land as well as to irrigate dry land.[33]
It would seem that Lee was engaging more in the theory of these
mechanics than in their actual practice, for few farmers would or

could ever go to the trouble or expense of using engines, although it would be within the reach of many to set up hydraulic rams.

Lee was much concerned with getting farmers interested in what elements were in their soil and what were lacking; and for that reason, he was always ready to analyze samples that were sent to him, free of charge.[34] With land cheap and labor high, there was always the temptation to leech the fertility and move on.[35] This led to soil exhaustion, which Lee held was one of the greatest menaces to the nation; it was "not sectional, but American in character."[36] "The continued fruitfulness of the earth," he held, was "an interest far greater and more enduring than any form of government."[37] The two most prominent features of American agriculture were "productiveness of crops and destructiveness of soil."[38]

Somewhat pessimistic with his exaggeration and guessing, which he severely condemned in others, Lee said that the country was losing annually $300 million "by the needless destruction of its agricultural resources."[39] For making a statement before a Georgia gathering that the Southern "system of tillage and husbandry" had worn out over a hundred million acres of land, a bully threatened him with a bowie knife for insulting the South.[40]

How absurd it was for people to keep taking the fertility out of the soil and putting nothing back. Lee likened it to a person who continued to check out his banking account and put nothing back until it was exhausted. "We are forever drawing on the soil for food, raiment and every species of wealth," Lee said, "and think it a great hardship if required to deposit in return one-tenth the amount we receive."[41] The menace was much wider than to the individual alone. He hoped to be pardoned "if it be a little out of place to urge the wickedness by seeking wealth by seizing and selling the natural capacity of the soil to support mankind."[42] It was "a momentous theme," as lightly as it seemed that many considered it, "for the soil we now scourge will one day refuse us both bread and meat, unless we treat it better."[43] In fact, it was flying into the face of the Lord, for could "a Christian people show any moral right to render ten or twenty millions of acres in Georgia less productive than they found them?" If so, then the next generation would have the right to "complete the work of destruction."[44] In 1857, he said, "At the present rate, it will not be more than fifty or seventy-five years before the Southern States are one vast desert."[45]

Lee never ceased preaching this sermon, even with some headway that he and others were making. In 1872, he said that for fifty years he had been denying "the right of any generation of farmers to leave the soil that must feed and clothe the next and all succeeding generations any poorer than they found it."[46] But about the same time, in a different mood and admitting progress, he said that "we have six hundred million acres of the best farming and planting lands that the sun shines upon."[47] Lee bragged that America was "the only country on the globe where every human being has enough to eat," and with the millions "coming here for bread," people must quit impoverishing "ninety-nine acres in a hundred of all that we cultivate."[48] And any people with one eye open should be ready to consider the question: "How they can best feed the land that feeds them."[49] The first lesson in husbandry was "to learn how to husband all the elements of fertility."[50]

Lee did not preach doom without providing the remedies for preventing it—fertilizers, soil builders, nature's rhythm, and so on. If let alone, nature herself would not abuse the land; and, if man would desist for a time by letting the land lie fallow, nature would restore the fertility. "Grass, sedge, briars, wild plums, sumac, and forest trees," Lee said, were "nature's agents for regaining lost fertility on lands which have long been tilled and turned out as an open common."[51] In reclaiming waste land, especially eroded red hills on which no vegetation could secure a foothold, the pine tree was man's greatest friend. The pine was an evergreen and grew the year round—Lee was writing for his Northern audience when he said: "Here is the first letter of the alphabet in the science and the philosophy in accumulating bread and meat in the surface soil. Study the ways of Providence; wisely initiate His example, and a barren plain can be converted into a fruitful garden."[52]

In Lee's most ambitious article on soils, which he entitled "The Study of Soils," and published in the Patent Office report for 1850, extending it to 56 pages and studding it with chemical analyses, formulas, and tables, he went into the origin of soils and fertility. He must have enjoyed especially writing about mold: "If the forest-leaves that annually fall to the ground did not rot, and become ultimately dissolved and return to their original elements, they would soon accumulate to the depth of many feet. The progress of decay in all vegetables and animals and their products, is governed by fixed natural laws; and these organized bodies decompose in that way which will best promote the growth of new generations of living beings. *Mould* is the half-way house between the living and the dead in the organic and inorganic worlds. It covers the surface of islands and continents like a carpet, and is a treasure of inestimable value when properly used by the husbandman. The fertilizing power of mould depends partly on its peculiar mechanical and porous structure, and partly on its chemical composition."[53]

Lee was tenacious in his views on fertilizers and their analyses; he could be made to depart from his customary gentility toward everyone on this subject more than on any other. He fell out temporarily with John J. Thomas, a former editor of the *Genesee Farmer*, over the analyses of certain fertilizers. In the dispute Lee used such vigorous expressions as "errors," "absurdities," "humbugs," and "gross perversions." And Thomas came back with the demand that Lee "retract his former censorious and uncalled for remarks."[54]

Lee did not have to dispute with anyone over the value of lime in aiding plant growth on certain lands not blessed with a limestone base. He devoted whole lectures and articles to the one subject, lime: its value, where it was needed, and how to apply it. The great civilizations of the past, he said, were based on the fertility that lime gave their lands. Both Babylon and Thebes grew to their greatness in limestone valleys. In Egypt the "lime rock that furnished material for the pyramids gave *wheat* to the laborers." Greece and Rome knew the value of lime and used it in their agriculture.[55] More poor land had "been renovated by the aid of lime than by the use of any other single elementary substance," he said.[56] He could not understand why the railroads did not give low rates to the freightage of lime, for thereby they would greatly increase the farm products they would carry to market: "Now, their agricultural freight is all in one direction—from the robbed land to the robber markets."[57]

Peruvian guano, the droppings of birds mined from the islands and coasts of the mainland, became a rage among many farmers in antebellum times; but Lee did not join in. He had analyzed this fertilizer and found that it lacked several important chemicals good for plant growth. He wanted to know why people thought bird manure, brought from thousands of miles away was so much better than cow manure, which could be garnered on every farm. Homemade manures were vastly cheaper and better, and especially so when used with South Carolina phosphates, "wood ashes and clover or cow peas," for "one can fertilize a field much cheaper than to buy Peruvian Guano."[58]

It was only a step from imported Peruvian guano to homemade commercial fertilizers, which were coming much into vogue in antebellum times. Lee admitted that there was much value in them, if properly made by responsible companies; and he warned farmers and planters to be wary of anything on the market under the name of a commercial fertilizer. Much money, he said, was being thrown away on *"common dirt,* bought under some high-sounding name as a valuable fertilizer."[59] Later in life, when commercial fertilizers were being marketed more responsibly, Lee favored the standard brands over homemade manures, though not denying the value of the latter.[60]

Saltpeter (niter), which was an element of gunpowder, was also a valuable fertilizer and was widely marketed in a generation later than Lee's as Chile sodium nitrate. Lee highly favored it, saying that it had "precisely the same force to make corn and wheat grow" that it had "to drive a cannon-ball through the air."[61] There were also many other easily obtained materials which he recommended as a fertilizer. Salt was valuable[62] and so was soot; the latter might be "used with great benefit for wheat."[63] Even woolen rags and bones were good. To prepare bones, one should dissolve them in sulphuric acid, or what was cheaper, wood ashes. Also, they could be ground up and sprinkled over the land. They would, of course, have to dry before they could be pulverized, and drying could be easily done in a dry kiln. He said this method was much used in England.[64] There were other fertilizers which Lee advocated, including certain grasses and leguminous plants, which he discussed in dealing with livestock.

# ❧ VIII ❧

# Rural Diversification—
# Grasses and Livestock

CROP DIVERSIFICATION was an old and widely discussed subject in antebellum times, as, indeed, later, and had Lee not been in the forefront, in addition to being an unusual agriculturist, he would have been a peculiar if not eccentric one. He stood for not only the usual agricultural diversification of crops, but also for doing everything helpful that could be done to agriculture. And always giving credit to the laws of Nature and Nature's God, as science revealed them he found that diversification was one such law. Just as Nature decreed it, "the public good" demanded it.[1] "Nature is the cultivator's best teacher," he said. "The same Divine law that covers a plantation with many kinds of forest trees, adapts it to the economical production of different crops, whose aggregate value is considerably larger than that of any one crop, for a series of years possibly can be."[2]

Of course, cotton was the crop associated in the public imagination with the South, sugar was less so, and corn even less although it was more widespread than cotton and in its aggregate production was greater than in any other section of the nation. And as Lee had much to say about cotton when he discussed slavery, he was always ready to suggest new and even exotic crops that would lend not only variety to country life but also would provide a valuable financial income. For instance, there was the fig, which grew well in the South; all it needed was to be promoted. Using rhyming words, employed not wholly for literary effect, he said, "There is a fig and a pig now in the cotton States that has yet to be developed."[3] He predicted that figs some day would be "numbered among the staples of the Southern States."[4]

Although other agricultural leaders had advocated olives for the South, had seen them introduced, and had recorded their failure, Lee in his old age seems never to have known this, or had

67

thought that a new attempt should be made. Much of the South, he said, was superior to the olive area of France.[5] He believed that the "Southern farmer with an olive grove in bearing will get six times as much per gallon for oil from olives, pure and ripe, as for the juice of his best grapes."[6] But his reference to grapes was not to downgrade grape culture, for most of his life he argued for vineyards and wine making. Wine made from the indigenous scuppernong was especially to be encouraged; adding a little sugar to tone down its tartness should not be considered an adulteration.[7] As Lee studied the climate and topography of the United States and noted its great varieties, he felt sure that the tea business could well be introduced, and he believed that the South offered the greatest opportunities for its introduction. "I feel warranted in expressing the opinion," he said, "that the time is not far removed when Southern enterprise and field hands will excel the Chinese as much in the simple operations of picking and curing tea leaves, and growing the trees, as they now do in growing, picking, and ginning cotton."[8]

Southerners did not need to be told to raise sweet potatoes, for they had long enjoyed this delectable food dug from their fields and gardens; but they had discovered no way to keep them long enough from rotting so that they might be marketed during the winter months. The fertile mind of Lee suggested a way to save them. By peeling and drying them, the planter could market them far away. Furthermore, they were rich in starch and sugar; when properly dried, they could be ground into a powder, and by mixing the powder with flour or corn meal, Lee said that farmers would "improve our daily bread."[9]

When Lee was a boy back in New York State, he helped set out apple trees, which were still bearing seventy years later.[10] Here was a hint for the South to develop its apple orchards into a big business. He advocated this in his old age and probably knew of the beginnings of the apple business, which in Virginia was later to grow to vast proportions. He said that at this time there would soon be 400 million people living in Europe, and there was a market for apples as well as for cotton.[11]

One new crop that Lee promoted, as, indeed, did many other agricultural leaders, a crop that seized the imagination of people and produced a veritable storm of excitement, was sorghum—sorgho sucre, kaoliang, Chinese sugarcane, botanically sorghum succharatum, as it was variously called. Agricultural journals and newspapers in all parts of the country published articles on this new plant.

"We have not room for a tithe of these articles," said Lee, in the *Southern Cultivator,* in 1857, "nor can we publish half of the private letters we have received on the subject during the past two months."[12] A French consul in China had seen it growing there in the northern part of the empire and, being much impressed by its possibilities, he introduced it in France in 1851. Three years later, D. J. Browne of the Patent Office, on a mission in Europe to collect agricultural information and new plants, saw this crop growing in France. He secured about 200 pounds of the seed, which was distributed to congressmen. The next year, on a similar mission, Browne secured in France several bushels of sorghum seed, which was said to have come from South Africa, where fifteen varieties of this plant had been found growing.[13] About this time a Boston seedman imported some of the sorghum seed.

Dennis Redmond, an associate of Lee's on the *Southern Cultivator,* secured some seed from the Boston firm and planted them early in the spring of 1855. Being quite successful in this venture, he soon began distributing some of the seed over the South and always insisted that he was responsible for getting the sorghum crop started in the Southern states.[14] Naturally, Lee began experimenting with this new crop. He made some "light colored, thick and very sweet" syrup from the juice, a sample of which he gave to the editor of the Augusta *Constitutionalist;* Lee predicted that he could further rectify the juice and produce sugar.[15] In 1858 he bought seed enough to set out a hundred acres and gave most of the seed away to various planters over Georgia and the South. Keeping enough for his own use, he made syrup and enthusiastically declared that it was "after keeping five months, nearly as light colored as the best honey, and as thick and palatable."[16] An instrument called the sorgho saccharometer was soon invented to show the concentration of sugar in the syrup.[17]

It was not long before Lee came to the conclusion that sugar could not be made from the sorghum cane, since no way had been found to produce granulation; and he advised that the best thing to do was to boil the juice and make it into molasses or syrup.[18] He did admit that "one of the most profitable products" was to let it ferment into alcohol, "but we sincerely hope that it will be manufactured entirely for mechanical uses, as there is sufficient 'blue ruin' in the world today."[19]

Of course, such crops as corn, cotton, and wheat were not new to the South; but Lee suggested that the cross-pollination of varieties of corn from the North with that from the South might pro-

duce a valuable new variety;[20] that there might be improvements
in cotton seed; and that the growth of wheat should be greatly
extended. He said that he had long lived "in the best wheat-growing
district in the Union," Monroe County, New York, and that he had
devoted years of study and observation on all the influences of soil
and climate and other factors in its growth.[21] He knew of several
New York counties, each of which raised more wheat than the
whole state of Georgia.[22] It was time that Georgia and the rest of
the South outside the coastal plains go more into wheat raising.
He gave his readers sufficient information on how it could best be
done.[23] Much of the South in both soil and climate was well suited
to wheat. He knew a planter near Augusta, Georgia, who after
raising a wheat crop, sowing it in December and harvesting it in
May, had seven months remaining in which to plant other crops.
Nothing like that could ever be done in New York.[24]

Wheat and corn suggested their disposal by selling, feeding, or
eating; and the eating of these grains called first for grinding them.
Why pay one-eighth of a bushel of wheat to get seven-eighths of it
ground? That belonged to the dark ages. There were many streams
in the upper South which provided power for running grist mills;
$3,000 would be sufficient to provide a mill that would grind 200
bushels a day.[25] To the large planter this might seem enticing, but
Lee did not go into the fact that the small wheat farmer, unable to
erect a mill and not needing it but for a day or two out of a year,
could not get out of paying the one-eighth toll, so reminiscent of
the dark ages.

To provide larger agricultural yields from the land, Lee did not
neglect to instruct farmers on the best methods. Among the various
aids he advocated, as did other agricultural leaders of his day, was
deep subsoil plowing. "It is not enough to scratch the ground as
pigeons do to garner beechnuts," Lee remarked in 1854;[26] but
some years later he warned farmers not to plow too deep and turn
up too much subsoil.[27] Another suggestion he made was to quit
pulling fodder: "We should as soon pull leaves from field-peas as
from corn stalks."[28] Giving a good reason why, he said that "fodder
costs at least five times more per 1000 lbs., than to raise a mixture
of orchard grass and clover."[29]

In his campaign for diversification on farms and plantations, Lee
looked to the rivers and creeks to afford a small fishing industry; he
would anticipate a movement that caught on a little later by suggesting that where there were no rivers or creeks large enough for
fish, "artificial ponds" might be developed. "The time is not far

remote," he said in 1853, "when the cultivation of fish will become an important branch of rural industry in the United States."[30]

Forests were a distinct asset to the rural economy, and any farm or plantation devoid of trees was much the poorer. Hoping to check that frontier aversion to trees, the "felling of natural forests," Lee admonished the people, had "already been carried too far in many portions of the United States."[31] "No people," he said, "have more reason than the Americans to admire 'the goodliness of trees:' and yet in no country are they more rudely assailed as the enemies of cultivation, and objects worthy of extermination by the ruthless axe and consuming fire." In 1852, he predicted that "forest culture . . . as a science and an art" would in time become popular in the United States.[32]

Lee could hardly restrain himself in writing about the Southern climate, about the "Sunny South," which gave it a great advantage in agriculture over the North. He said that reliable statistics showed that "the citizens of Georgia are worth *per capita* 150 per cent more than the citizens of the State of New York. We do not say that the citizens of our adopted state are more industrious, intelligent, or more economical than those of our native State; but we do say that one can raise a crop of wheat and one of corn in succession on the same land in Georgia, in the time consumed in the growth of either crop in New York."[33] He said that even as far north as Virginia and Maryland, climate gave those states a 50 percent advantage over Oneida County in New York.[34]

Having long studied the effects of climate and weather on agriculture and the growth of plants and vegetation, Lee ran a series of articles in the *Cultivator & Country Gentleman* under the title of "Agricultural Climatology."[35] Relating climate to the great plains east of the Rocky Mountains, he predicted that the region would become a great wheat-raising country.[36] Then, discussing the effect of a humid climate on slow growth, he thus accounted for the big trees of California, as well as the production on the humid Pacific of turnips weighing 100 pounds, cabbages weighing as much as 56 pounds a head, sugar beets weighing 47 pounds, and onions of 21 pounds each. Then, continuing to relate time to growth, he said: "This is, perhaps, the most potent element in vegetable organization; and to gain the full benefit of this, slow growth in any equitable temperature . . . is a matter of great practical importance. To woody tissue time gives increased strength, solidity and value. To the cane it secures more sugar, matures joints and buds; to the cotton plant, more seed and lint; to wheat, rye, barley, oats, timothy

and other grasses, more seed and nutriment; and to the beet and tur-
nip, larger roots; the potato a larger tuber, the cabbage a larger
head, and to the clover plant an expansion that spreads its rich
herbage, from one root, over a surface of 81 square feet."[37]

Climate affected not only plant growth but also people. As Lee
studied history he found that the great men of past ages had been
born in warm climates. With this knowledge in mind, he advised
Northerners to migrate to the South not in spite of the climate but
because of it.[38] In explaining why Southerners were different from
Northerners, a difference in Southerners which he liked, he said,
"Southern ideas, so far as they are peculiar, are the offspring of
climate quite as much as buzzards, alligators and cotton."[39] And
when he was eighty-five years old, climate was as much food for him
as meat and drink, for he said, "as a plain working farmer, I am able
to transform bright and beautiful sunshine into thought into my
own brain in ten days time."[40]

Diversification: it was a great thought in Lee's mind. The Great
Divine set up its laws; man must not try to evade them, for if he
did so, it was to his own hurt. Lee was not blazing new trails in his
advocacy of diversification. His contribution to the program was
the frequency with which he set it forth. His name carried great
respect throughout the South, and the fluidity and earnestness with
which he wrote and the conviction and authority with which he
spoke commanded ready readers and listeners.

The lower South was not a grass country, and, hence, it had never
gone into the livestock industry either in dairying or in raising
its draft horses and mules. Realizing the handicaps that had prevented these developments, Lee still believed that the South could
do much more along these lines than it had been doing.

One of Lee's most constant interests was grasses; even in an article
as far away from the subject as "Agricultural Engineering," he
could not refrain from bringing in his advocacy of grass culture.[41]
He liked to say "all flesh is grass."[42] He was historically minded in
almost anything about which he was writing; how easily it was for
him to relate grass to history! The "old Agricultural Romans conquered the ancient world, mainly through the inherent, God-given
power of grass," he said.[43] And without grass there would have been
no England: "It was the intrinsic value of grass and hay that made
England the greatest commercial nation in the world."[44] The "very
best aristocratic blood in man and beast," he said, was developed
out of grasses. Writing in 1860, he declared that the British aristocracy, which ruled most of the civilized world, owed "their power to

grass." The American people by their purchase of livestock from the British Isles for breeding purposes paid "both homage and tribute to the wise grass culture of the only people in the Old World who speak our mother tongue." He intended to show his readers and listeners "the inseparable connection that subsists between grass and that calm, cool, far-seeing and most perfect brain-work of which man is capable."[45]

Although Lee would not be foolish enough to "let the country go to grass" by emphasizing grass culture to the exclusion of the plow economy (which was, after all, the broad basis of his farm program) and, thereby, make nomads out of the American people, yet he could in his enthusiasm for grass almost seem to make it so. In 1869, when he was living in East Tennessee, Lee said, "I can raise 1000 pounds of hay as easily as 100 pounds of corn" and the hay brought twelve times as much money as the corn. "Grass culture and stock-raising require very little labor, save all fields from washing, improve land, give sure crops, and large profits. Southern farmers plow too much surface, and [in] every way overdo tillage, as though it was the beginning, middle and end of all good husbandry. . . . Will not money from plants that grow ten or twenty years with no use of plow or hoe, be as acceptable as that derived from wearing out already impoverished plantations?"[46] When he was writing in this frame of mind, he could easily say, "if either department of rural industry should predominate over the other, planting should yield to grazing; in order to save the land from the scourge of the plow."[47]

Farmers should save the seeds of their various grasses either for sale or for wider sowing. Lee always harvested his grass seed, and most of his seed he gave away in support of his campaign.[48] The seed he kept, he liked the feel of, and as he said, "I sow all seeds by hand, and use no hand but my own."[49] By continually pounding away in his writings on grasses, he aroused much interest and considerable curiosity about certain grasses that farmers found growing on their lands. They frequently sent samples to him with the request that he identify them. Lee was always glad to do so, without charge. He liked to consider himself an authority on grasses, as he should have been after all of his advocacy of spreading them. In 1855, he said that he had "a collection of twenty-eight genera, and sixty-two species of the most common grasses grown" in the Northern states and in Great Britain.[50] He stated that twenty of the species grown in England would be suitable for the South.[51]

Lee always liked to use homely similes and comparisons. "One may plant a gold dollar in the best soil," he said, "and never obtain a dime or a cent for his labor and capital. A dollar's worth of grass seed may be multiplied into thousands of dollars in a soil where gold and silver fail."[52] And how he liked to tell how nature transformed one thing into another—ultimately into flesh and blood and into money: "What is easier than to change rain and dew into grass and hay, and both into money?"[53] When he was nearing the end of his life, he said that for forty years he and his family had lived on farms and supported themselves there, "never going into debt a dollar to a store, bank or mechanic," living from the economy of clover and grasses, feeding two or more cows, making skimmed milk for the pigs and butter for his table and the market.[54] That was the ideal life without too much use of the plow, but in plow culture, when "laying by time" came in the corn field, there was still a chance to go to grass by sowing in the field, clover, orchard grass, timothy, and herd grass, which could then be grazed after the corn had been gathered.[55]

There were many things in favor of a grass culture: "Bad laboring people may steal corn from horses and mules when fed; but how can they steal green clover and orchard grass in a pasture?"[56] Milk was seven-eighths water; hence green grass, rather than dry hay, would best yield the sugar, butter, and curds found in milk. Even in winter, when there would be no green grass, steaming or cooking hay would give it some of the qualities of green grass.[57]

Lee found it necessary to argue constantly the value of grasses to overcome the hostility of the ordinary farmer against grass as being a pest, and the defeatism of others who, realizing the value of grasses, held that the South was too hot and most of the land too high and dry. Lee countered by saying that on his Georgia farm, he mowed his Bermuda grass three times a year and got three tons of hay at each mowing. "Let planters try half as hard," he said, "to make grass grow as they are to destroy it, and no one will need dried and baled grass imported from Yankee land." In 1861, he noted that five years previously he had sown alfalfa and Bermuda grass on a field and they had been flourishing ever since.[58]

Although for years in the antebellum South the prayer had gone up from many agricultural leaders for a perennial grass that would grow there, still there was much opposition to grass, for, like weeds, it got in the way of the plow culture. In a lecture in Burke County, Georgia, in the 1850s, Lee was laughed at for telling how to develop grasses when his listeners had all their lives been trying to get rid of the pests.[59] Lee tried to impress on people that grasses conserved the chemicals of the soil, whereas tillage leeched them out.[60] And when it came to "resting land" or letting the land lie fallow, too many farmers let it grow up "in weeds, sedge, briars and bushes, that will support good clover and orchard grass, and most profitable farm stock."[61] Late in life he was glad to say, "Fighting grass with something like fanatical zeal has had its day."[62]

Of course, Lee did not try to introduce the sixty-two species of grass that he said he had in his "grass museum"; he had his specialties for which he argued. Predicting in 1871 a great grazing future for the South, he recommended "orchard grass, tall oat grass, redtop, timothy, Kentucky blue brass and meadow fescue."[63] Sometimes he added to this list Texas mesquite, ryegrass, and herd grass. He included in his grazing fields red and white clover, which, of course, were not grasses. His enthusiasm for Bermuda grass varied and practically vanished. Finally he put it at the bottom of the list.[64] "Weeds are better than no vegetation," he said, "but rich clover and English grasses are far better than weeds and briars. Even Bermuda grass is better than a crop of weeds."[65]

Lee long centered his attention on Kentucky bluegrass. He said that Henry Clay told him that it was indigenous to America. Lee said he was interested in testing its vitality in comparison with other grasses.[66] He was no defeatist on making it grow in the South, although he knew that it needed a limestone soil, as was shown by how it thrived in the limestone regions of Kentucky, Tennessee, and Indiana. But he had proof that it could be made to grow and continue further southward. In 1862, he said a lady from Greene County, Georgia, presented him a specimen from a field which her father had planted fifty years ago, and it was still thriving. Kentucky and Tennessee wagoners coming through Clarke County, Georgia, and adjoining counties, camping over night at spring sites (including one on Lee's farm), scattered bluegrass seeds from hay which they had brought along. Lee said that these seeds sprouted and the grass had become so hardy that "they have held the ground against broomsedge, briars, weeds and all bushes, for thirty years."[67]

Yet Lee said that bluegrass was more likely to fail than any other grass; but if the soil should be enriched with limestone and otherwise prepared, bluegrass could be made to do as well in Georgia as in Kentucky.[68] With a process that he had developed, he said that he had produced from one seed 1,000 separate plants, which were growing on his farm—the "one seed planted only eight months ago." He was thinking of having the process patented, "more to protect the public interest than our own."[69]

Herd grass made excellent hay and grazing, Lee said, and it had the advantage of thriving on land too wet for farming—it made an excellent meadow.[70]

Lee tried to exploit mesquite grass (*Holcus lamatus*) to a little extent and without much success. To get seed, he sent to Texas where it grew wild. He said it was "one of the poorest of English grasses," which had been brought by Catholic missionaries to Mexico and then into Texas.[71]

Lee did not set himself up to be a livestock expert, and he did not have a great deal to say about it, but he could hardly have emphasized grass culture without having livestock on the end of it. But anything in which he was even mildly interested had to come under his close and long observation and involved some experimentation. In his New York days he had been associated with dairies to some extent, and he began early the study of "animal physiology." He was soon telling people how best to take care of domestic animals, how and what to feed them, the "importance of warmth and quiet," and how calves should be raised on whey.[72] He warned against letting cows eat acorns, as his cows had been doing for a time, for thereby they gave less milk.[73] Those who wanted to keep up the strain should keep herd books and never go out of them for blood. That was "breeding in the line."[74]

Lee wanted the South to engage more in dairy farming; it was part of his grass culture program and, of course, it was a fine example of diversification. This would lead to a butter and cheese industry. In 1848, he said that there were only two cheese dairies in Georgia, and both of them were run by New England families."[75] And there was "not a State in the Union in which both cheese and butter" could not be made.[76]

Instead of having well-developed dairy farms with fenced pastures, most of the owners of livestock in the South allowed it to run loose on the free range, a system that belonged "to a semi-savage era." "My stock," said Lee, "have no more right to invade my neighbor's unfenced crops than I have to burn his barn or house."[77] It was too great a burden for the farmer to fence his fields "against small shad bellied swine, jumping sheep, and unruly cattle." It taxed "farm land more than the use of it" was worth.[78] Lee realized how at one time, in the early history of the country, the free range was logical and practically necessary when fields were few and the open range was vast. But now, the situation was reversed. Livestock should be fenced and fields left open. Writing in 1855, he said, "Our present system of fencing against live stock, of turning them out to shirk for themselves, and too often [to] steal their living or starve, is bad every way; and it belongs legitimately, to the dark ages of semi-savage life; . . . let our stock be fenced in, rather than our corn and cotton. . . . Why protect by law an hundred dollars worth of property in hogs at a cost of one thousand dollars worth of fence?"[79] With all of the aid of other agricultural leaders who thought the same, still Lee made little headway.

Of course no improved or registered stock could be allowed to run loose, to mix with scrawny cattle, mangy sheep, and razorback hogs, and to produce like offspring and contract their diseases. Lee's inquisitive mind did not lead him into the field of veterinary medicine, though he did speculate on the origin of hog cholera as being connected with the importation of highbred swine,[80] and he studied horses sufficiently to learn that epizootic influenza was brought in from Canada and to discuss its treatment.[81]

The livestock that engaged Lee's interest most were sheep. He preached sheep to Southerners at every opportunity; he could not understand why they did not go in more for sheep, for sheep could be raised much more cheaply in the South than in the North. In 1847, he reported that in "a week's travel [in Georgia] I have seen but one sheep; while of dogs the supply is everywhere most abundant."[82] The next year, having learned some more about livestock in Georgia, he wrote, "There are, I believe, more goats than sheep in Georgia, and more dogs than goats and sheep put together."[83] Lee's aversion to dogs, of course, was based on their sheep-killing instinct; and in later life he advocated curbing their "abundance" by taxing them. He thought that one reason why a sheep was "the most neglected animal at the South" was the belief Southerners had that wool came into competition with the great staple of cotton. He quoted John Randolph of Roanoke as having said that he would go ten rods to kick a sheep; Lee concluded that cotton lands would have to wear out "before any planters of the South will abandon cotton culture and dog husbandry for woolgrowing."[84] If he could not reduce the cotton crop to increase wool growing by ordinary reasoning, he thought he had a point, which he did not push, by showing how the two "crops" were interrelated: cotton produced seeds, and sheep would eat them, especially if the seeds were de-hulled—and he said that there was a machine that would do this.[85] After the war when the slaves were free and labor was hard to get, Lee felt that it would be much easier to find shepherds than field hands.[86] He concluded that merinos were the best sheep for the South and that sheep were better than gold: "Gold in the earth never multiplies its particles, nor increases in weight. Sheep multiply rapidly, and supply both meat and raiment at a very small cost, if properly managed."[87]

# ❧ IX ❧

# Slavery and the Foreign Slave Trade

TO LEE'S Northern acquaintances and even to some of his Southern friends, it seemed strange that he should have entered into a strong defense of slavery and of the foreign slave trade. But the more one dips into his fundamental character, it does not seem very strange. He was old enough to become acquainted with slavery in his native New York before it was abolished there, and he saw the effect that freedom had had on the freedmen. He thought it made them shiftless and lazy and added nothing to the betterment of society in the community in any way.[1] And despite his temporary explosion on social betterment in connection with his report on agriculture in the New York legislature, he was conservative by nature and believed in the status quo except in his agricultural program.

True to his scientific nature, Lee saw slavery as a sort of outgrowth of natural laws, but he did not bring the Great Divine into the picture as he generally did in discussing the forces that direct man's life. "Soon the 'sin of slavery' will be entirely forgotten," he said, "as the laws of climate and of human industry are studied and understood."[2] God did not enter into his argument, and he never put forth the Biblical defense of slavery. Slavery was "pre-eminently a labor question."[3] It was not a moral question; it was an economic one. Least of all should it be thrust into politics and made a political football, eventually to disrupt the nation. Somehow he seemed to see, in 1849, "that the miserable political humbug about slavery, is slowly becoming less potent for mischief."[4]

With diversification of agriculture ever present in his mind, still Lee saw the South wedded to cotton as a main widespread crop, forced on the South by natural law and world demand. Supposedly referring to changing styles of dress for women, he said, "If all the women in the world insist on having twice as many yards in each calico dress as formerly, and twice as many dresses in the course of a year," who, then, but slaves would raise all the cotton needed "to supply this incalculable demand?"[5]

What was more, cotton was the wave of the future in more ways than in dressing women according to prevailing styles. The population of the world was great and growing both in numbers and in the uses brought on by civilization. The South raised the world's supply of cotton and England manufactured most of it to clothe the world, despite the attempts of the national government to build up the New England cotton mills through tariffs on English manufactures; and the South would continue to supply that cotton despite England's attempts to develop cotton fields in India and in the valley of the Niger in Africa. Lee warned England that she would have to be more restrained in her opposition to slavery and slave labor, and she was already beginning "to make her necessity a national virtue and to speak more respectfully of the kind of labor which produces her cotton. . . . In a word, people are not apt to quarrel long and earnestly with their bread and butter, and look with jaundiced eyes at the source of their wealth when once understood."[6]

Slavery was a civilizing influence and made for a perfect society, Lee said. It gave the slaveholder that leisure necessary to enable him "to cultivate his thinking facilities and improve his moral perception."[7] The great leaders of America's past had been born or had grown up in a slave society, not only in the South, but many of them, in the North. Naming a few in New York, he mentioned Alexander Hamilton, John Jay, DeWitt Clinton, the Livingstons, and others. And as for those born in the South, who could forget George Washington, Thomas Jefferson, James Madison, and a galaxy of others?[8]

Slavery was not only the making of the great planter aristocracy, it was the very essence of the slave himself, for if he were not a slave in the South he would be a slave or worse in Africa. "Compare the condition now of the natives in Africa with the negroes of the South," Lee said, "and every one must see that the latter have gained immensely by being transplanted from a land where civilization has not advanced one inch in four thousand years to the heart of a Christian nation."[9] Furthermore, he thought that it was "not wrong to give work to a black man, restrain his vices, cultivate his morals, and teach him at once the art of agriculture, or some mechanical trade, and the humanizing, elevating principles of the Christian religion."[10] Lee even had visions that in the course of time "we can send colored planters to Africa, every way qualified to civilize and christianize the natives of that country if it be possible."[11]

As slaves developed in skills and industry some of them might look forward to becoming freemen. And as an example, Lee cited the case of a Negro in Augusta, who paid his master $2,000 for his freedom and that of his wife and two children. Also, Lee knew of a slave in Augusta who, rather than buy his own freedom for $1,700 bought two slaves for $1,300 and received the return from their employment. (Of course, this situation was unusual and was outside the laws governing slavery in the South.) But, as Lee noted, slaves in towns buying time from their masters for $75 or $80 a year, earned for themselves as much as $1,000 a year driving drays, working as carpenters, and doing odd jobs. Thus, they accumulated money to buy their freedom or to spend their money as they pleased.[12]

There was another side to slavery as Lee saw it in operation in Augusta. Free Negroes who engaged in crime might receive as punishment the penalty of being sold into slavery for a term of years, not more than two for the first offense, and for life if the offense should be repeated.[13] This law was passed in December, 1859, and created a considerable stir in Augusta and, doubtless, in other parts of the state. Two prominent Augustans condemned it severely, prompting Lee to come to the defense of the legislature. He said there was nothing reprehensible in putting a few free Negroes into slavery, their being among four million other slaves in the South; slavery was a good school in which to elevate Negroes out of a life of crime. At this time, in addition to the two or three positions he generally held, Lee was the agricultural editor of the *Southern Field and Fireside*, published in Augusta by James Gardner.

In his dispute with the Augustans, Lee had been severe to such an extent that one of them wrote to Gardner, complaining about Lee's language. Gardner wrote a long letter apologizing for what Lee had said, explaining that when he had employed Lee as agricultural editor on the *Southern Field and Fireside*, he had had an understanding with him that he was not to discuss politics or slavery. Lee replied to Gardner, his employer, in good spirit, as might be expected, and said that in the very first number of the journal he had criticized the Patent Office for publishing statistics that were biased against the slave agricultural system in the South, and that in other instances he had brought up the subject of slavery. He cited that in his severe criticism of Helper's book *Impending Crisis*, a copy of which Gardner had loaned him, he had shown how Helper had manipulated his statistics against the South. Lee said further

that there had been no written rules about what he should not say
in his editorials and articles, that it had all been in oral conversa-
tions, and that he had never understood that slavery was not to be
discussed.[14] This little flurry did not interfere with the relationships
between Lee and Gardner, and Lee remained with the *Southern
Field and Fireside* until its first interruption in the second year of
the Civil War.

In his defense of slavery, Lee combated the assertion made then
and repeated for years thereafter that the slave system wore out the
land. "To assert that slave labor is incompatible with agricultural
progress and improvement; that it is at war with the natural resources of the soil, and carries with it wherever it goes either partial
or complete desolation; and that it is a law of its very existence to
make land worse than it found it, is to maintain that slavery is the
worst enemy to mankind which the world possesses." And he answered the question by merely asking it: "is it not ridiculous to contend that the use of slave labor compels planters to wear out their
plantations?" He struck home the conviction that soil exhaustion
was "not so much a Southern, as an American principle." Any
planter could "exhaust his cultivated fields as well by hirelings as
by bondsmen." The actual reason for soil exhaustion was the mad
scramble of people to make money, skimming the land of its fertility, putting nothing back, and moving on to new lands. The solution was "to *improve*—not impoverish" land. The old states were
not overstocked with slaves: "They have unimproved land enough
for thirty million laborers more than they have."[15] In his zeal to
emphasize an argument Lee found it easy to exaggerate.

Lee was much impressed by the large amounts of unused lands
in the South, most of it excellent for cotton, sometimes estimating
it at 100 million acres and at other times, at 300 million, depending
probably on how far west he let his imagination extend. If there
were laborers enough to work this land, he ventured to say that "the
South would increase beyond all calculation." There were not slaves
enough to fill this demand for labor; he reminded his readers that
the British were solving their labor demands in the West Indies by
bringing in laborers from India and from China. Lee argued that
these people could be brought into the South as apprentices for a
term of years, and when their time was up, they would be returned
as educated farmers, and a new supply could be imported. Thus the
South would profit, and so would India and China. This would
be a balance for the South as against the North, which was receiving
tens of thousands of immigrants to add to the labor supply and

wealth of that region.[16] The wily abolitionists were keeping these immigrants from coming South. "Great pains," he said, "have been taken to prejudice the hundreds of thousands of European laborers who have recently emigrated to the United States, against the South as a field for the successful exercise of their skill and industry." These Northerners spread the notion "that a white man needs an umbrella over his head while working in a cotton field, to lessen the depressing influence of solar heat."[17]

Lee's zeal for filling up these unimproved lands with fields of cotton led some Southerners to charge him with forgetting all his diversification ideas and running "into the popular breeze, or cotton mania, which all farmers know has ruined Middle Georgia." Writers were calling Georgia "the Empire State of the South!" Georgia was dependent for almost everything except for two or three articles with which to buy the rest, said a correspondent signing himself "L. J. S." "The Empire State of the South, indeed! ! ! it is all a humbug," he said.[18]

It was only a step from bringing in Chinese and Indians, for Lee to join the movement to reopen the foreign slave trade, which broke out prominently in the latter 1850s.[19] His first involvement centered on answering a pamphlet entitled *An Argument against the Policy of Reopening the African Slave Trade*, written by Robert G. Harper and published in Atlanta in 1858. This was only one of a number of pamphlets in a pamphlet war, one of which L. W. Spratt, editor of the Charleston *South Carolina Standard*, was publishing the same year; and others in South Carolina and elsewhere were writing and talking in favor of the movement.

Lee was set onto Harper by James Gardner, who was not only the founder of the *Southern Field and Fireside* in 1859, but also the proprietor of the Augusta *Constitutionalist*, and it was in this newspaper that Lee carried on his arguments.[20] With his accustomed vigor he began his campaign for reopening the African slave trade. Lee had several strings to his bow: Always for the common man, he argued for the many would-be slaveholders as against the few; the South needed a great labor supply to develop its vast resources; and the benighted Africans needed to be brought to a land of civilization and Christianity. He developed other arguments as needed. Those who opposed the foreign slave trade, Lee said, "admit the value of slavery in very small doses, taken by a few; but they seem to regard it as an exceedingly dangerous medicine, which will kill the masses if they touch it."[21] Slaveholding would greatly increase the economic and cultural level of the two-thirds of the population

in the South who owned no slaves: "Give the poor white people of
the South an opportunity to own, each, only two slaves, and they
will rise at once into a higher and nobler condition." Thus, they
would have "all those social and industrial advantages, which made
Washington so wise, so rich and so patriotic." [22] If Congress was as
intelligent as the many people who want slaves, it would repeal the
law against the importation of slaves from Africa. The only argument against it would be that slavery was immoral and should be
suppressed. "If slaveholding is right at all," Lee argued, "it is right
every inch of the way from Africa to Georgia." [23] His children and
his grandchildren should not be deprived of the right and opportunity to own slaves. [24]

The common man in the South could not buy slaves because the
price was too high; opening up the foreign slave trade would bring
the price down within his reach. Returning to the few against the
many, Lee held that if "the practice and the principle of slaveholding are worth anything, they must be as good for all as for a favored
few." [25] A "Georgian" answering Lee contended that the masses of
people did not want slaves, and if they did how could they buy
them and where could they work them, for some of them (especially
the "poor whites") had no land and little energy or inclination to
do anything for themselves. If they should be given slaves without
cost, they would soon sell them and waste the money. [26] Lee's arguments were being labeled as dangerous, for he was stirring up class
consciousness where none then existed. [27] "Conservatism" charged
him with suggesting that a party might grow up to vote against the
slaveholders. And this critic said that slave agriculture did wear out
the land. [28]

Lee was always impressed with the wasteland in the South that
could be put to good use. In 1859, he said that there were 100 million acres of such land, which raised "millions of bushels of frogs,
snakes, and alligators in undrained swamps, and chills and fever to
match; while in the wilderness of upland, other millions of wild
animals, and wild children, may also be reared." [29] Answering those
who were opposed to raising more cotton, Lee argued that with a
little training slaves could be put to watching sheep; the South
could easily support 100 million sheep. It would take 500,000 shepherds to look after them, and where could they come from except
from Africa? It might take 100,000 white men to act as head shepherds, but in that fact there would be employment for many poor
whites. [30] Here again, it appears that Lee was letting his enthusiasm
for sheep husbandry run away with his practical sense; but it must

always be remembered that he thought in terms of a longtime future—even as long as the next hundred years. The future possibilities in developing the South's natural wealth were incalculable. Its minerals had not even been touched and only awaited that labor which Africa could supply. "If I bring ten negroes from the valley of the Niger to till my land, I wrong no white man in Georgia, or elsewhere, a particle more than I should by bringing ten Germans from their fatherland to do the same thing."[31] Lee did not note that one would be a slave and the other a free man; but he never considered slavery anything more than a labor supply.

The world needed Southern cotton and the nation itself needed cotton and other products of slave labor: "If nobody but slaveholders consumed the products of slave labor, there would be no necessity of having more slaves; but to meet the wants of the abolitionists alone, will keep seven million slaves hard at work."[32] England in trying to suppress the foreign slave trade by using her fleet to intercept ships bringing slaves from Africa was guilty of hypocrisy, for at that very time she was promoting in the valley of the Niger and elsewhere in Africa the raising of cotton by slave labor much worse than what prevailed in America.[33] Lee admitted that the transportation of slaves from Africa to America, the so-called "middle passage," was bad, and that the fact that African chieftains made war against one another in order to capture slaves for sale to America, also, was bad; but Lee claimed that tribal wars would continue to provide slaves for the Arabs and that the "middle passage" could be made more humane if there should be expended for this purpose one-tenth of the money then being used to suppress the trade.[34] Furthermore, Lee would not let his humanitarian instincts lie unexpressed. It would be bringing the poor Africans to a civilized land where they could come under its influence, for in Africa there were "millions of slaves . . . owned by benighted, stupid negroes in a heathen land, who are worth next to nothing to themselves, their children and the world." America had an obligation to help in this humanitarian work, for "man is as much in duty bound to improve and cultivate his fellow-man as he is to cultivate and improve the ground."[35]

It therefore behooved Congress to repeal the prohibition of the foreign slave trade. Lee said that Americans could import "work horses, mules, oxen, or camels, or not to import, and not to work either, as he pleases."[36] Why should the importation of other forms of labor be prohibited? Every Southerner had a right to be a slave-owner, but how could he exercise that right with prices so high?

In prohibiting the foreign slave trade, "the Federal statesmen saw fit to deny . . . [people] the use of slave labor, precisely as some southern statesmen now deny more than five million of southern citizens any share in the direct profits of negro slavery."[37] If the foreign slave trade were allowed, the price of slaves would be brought down to where any Southerner who wanted a slave or two could make the purchase, for in Africa they could be bought from $10 to $15 each.[38] The North could import "the cheapest pauper and serf labor from any and all quarters of the world," while the South had to go to Virginia or Maryland and be forced to pay from $1,200 or $1,500 apiece for slaves. But the law, Lee said, "forbids me, and makes the act piracy, if I choose to go to Africa or Cuba, where I may, for the same money, buy ten or twelve slaves." This situation was "more tyrannical, odious, and oppressive" than the British tyranny against which the Americans fought their Revolution.[39] "If it is right to hold persons as slaves in Africa and America at all," Lee said, "then it cannot be worse to transport slaves from the Niger to Savannah, than from the Potomac to Savannah for sale, as is now done."[40] And as for the charges sometimes brought that Southern slave-owners considered it more lucrative to work cheap slaves to death and buy new ones than to work them less and let them live longer, Lee said the mere statement of such an absurd charge was answer enough.[41] Also, Lee did not let Southerners forget that in adding to the population by bringing in more African slaves, the South's representation in Congress would be increased by the operation of the constitutional three-fifths rule.

Lee's mind was not keen enough to sense that there was much more involved in the institution of slavery than its being a labor supply. Now and then he gave evidence of seeing a little more, as when he used the term "apprentice for life" instead of slave, and when he included a qualification in a statement he made in 1858 that it took him "some time to make up his judgment to the effect that negro labor, as it exists at the South, is, upon the whole a good thing."[42] Lee's position on slavery had come largely through his own thinking and observations; there is little reason to suppose that he was much influenced by the various books and pamphlets defending the slave institution, and certainly the era of arguments had largely run its course when in 1860 E. N. Elliott's *Cotton is King, and Pro-Slavery Arguments* was published.

Lee did not escape condemnation in the North, as, indeed, was the case of all who defended slavery, for up to this time he had lived most of his life in the North, and many of those who opposed his views thought that he should have known better. He was first attacked for saying that soil exhaustion was no more a Southern than a Northern problem, and that the slave was no more guilty of causing it than the hireling was. In 1849, while Lee was still editing the *Genesee Farmer*, one of his correspondents chided him saying that "you seem to be growing very *tender footed* on the subject of the 'peculiar institution,' " and affirming that slave agriculture did deplete the soil, he added, "The *pecooliar* institution must be abandoned, or new worlds discovered for its use." [43]

In 1854, Lee wrote an article for the April issue of the *Southern Cultivator* entitled "Hireling and Slave Labor," and in the following June issue he had another article, "Agricultural Apprentices and Laborers," in which he pulled together much of what he had been thinking and writing on the subject of slavery. These articles, especially the first one, set off a jeremiad against him in the Northern press. Obsessed with the conviction that the South had vast potentialities in millions of acres of unimproved lands and the natural resources that went with them, he believed that all that was necessary to bring about a realization was a labor supply. The Southern planters could not continue to buy slaves from the border slave states because prices were increasing and, as those states were becoming less wedded to slavery, there was danger that they might enter the Northern orbit. The North had its labor supply not only in its natives but also in a flood of immigrants who did not go South; in addition there was an inclination of white natives against working in the fields with slaves, though Lee on his District of Columbia farm was able to use white laborers, and he thought they performed better than slaves.[44]

As previously noted, he had been suggesting the possibility (taking a note from the British book) of bringing in Chinese coolies and laborers from India. But adding his own further solution against having these people remaining permanently in America, he proposed that they be returned to their native lands after a term of years as apprentices, Christianized, civilized, and ready to do for their countries more than all the missionaries would ever be able to accomplish.

But however much Lee toyed with the problem, he found himself always returning to the great labor reservoir in Africa. Now, in 1854, in these articles he was not advocating the African slave trade —he would get to that later. He was thinking how wonderful it would be to bring Africans out of the barbarism to American civilization through a system of apprenticeship. He wrote, "Labor appears to be the effective means appointed by Providence to transform a herd of ferocious cannibals, fearing neither God nor man, into the most upright, cultivated, industrious and useful specimens of humanity. Savages never lack liberty, about which narrow minds prate so much, yet they never cease to be savages till agricultural labor changes their *ferae naturae*. Until this great and radical change is fully consummated, involuntary work of some kind is indispensable, or man easily or naturally reverts like the former slaves in St. Domingo, back to idleness, vice, crime and hopeless brutality. To start him out of the degradation of the brute, and lift him into the sympathies of civilized life, there may be better things than the plow, the hoe, and the whip, but neither ancient art nor modern science has discovered them."[45]

Following up the attack made on him for his article "Hireling and Slave Labor" in the *Cultivator*, Lee in his June article "Agricultural Apprentices and Laborers" discussed further how Africans could be saved from their degradation. He said that he had learned from an article in *Blackwood's Magazine* how little children had run after missionaries with the hope that they might find out how white people tasted. History showed that "eating of missionaries is no feat on the part of savages; and if civilized man has a right to subdue, tame, teach and evangelize wild men, then we repeat what we said on a former occasion, the plow, the hoe, and the whip are the best known means to accomplish such purposes; they being essential parts of a preliminary education. It does not follow because apprentices are legally bound to serve and obey their masters, that they may be wronged, in any respect, with immunity. Harshness and cruelty to servants are as unnecessary and unprofitable as they are brutal and unchristian-like."[46] Here Lee was thinking and writing in terms not of permanent slavery but of apprenticeship. The plow, the hoe, and the whip were not to be used as permanent methods of controlling slaves but in cleansing them of their African barbarism and educating them into Southern slavery, or apprenticeship, after which they should be returned to Africa as missionaries to their own people.

The *Rochester Daily Democrat* led the pack that set out after Lee. It began the onset by printing a part of Lee's "Hireling and Slave Labor" article and inviting to it "the special attention of our readers, premising that by 'hireling labor' our learned contemporary means that which sectional prejudice and fanatacism, here at the North, usually calls 'free labor.' Hirelings, attend!"[47] Lee's article was "merely a small piece of toadyism—and we must say, about the *smallest* it was our good fortune ever to meet—designed to make capital for the *Southern* Cultivator."[48] He had "been trying to sell himself South for some time" and was guilty of "Treason to Freedom."[49] It accused him of having two policies concerning slavery, one for the *American* and the other for the *Southern Cultivator.*[50] What the *Democrat* wanted to know was this: "The question before the public mind is simply this, whether the editor of the *American* has any principles, and if so, what they are."[51] Then sarcastically it continued,

What system was it that robbed this poor fellow of all his previous earnings, and left him a beggar in the midst of his years? What system had driven him forth from his native spot, where he had earned a reputation, and a right to live, by mingling his own sweat with the soil on which God had given him birth; and thrust him with his wife and little ones into a strange land and unaccustomed modes of life and labor, *on peril of which is dearer to man than bread*—his manhood, his holiest affections, in a word, his freedom.[52]

In fact, it took a long time and much of the space in the *Democrat* to relieve it of its indignation over what Lee had done. A small but pertinent part of its long tirade against Lee came in these words:

At any other time we could laugh at the inanities of this disgusting piece of flunkyism; but at an hour like this, when the dark shadow of despotism is spreading itself all over the land and free men stand aghast at the insolent aggression of the slave-power, it is really no jesting matter, that there should be a man in our midst, who is thus playing into the hands of our enemies and dealing deadly stabs at the most fundamental and precious doctrines of human rights—and *that*, a man who professes to be the conductor of a free press and the instructor of free laborers. Nor can we forget, or remember without a blush of shame and indignation, that it is with leaven of this sort, he is filling the Rochester American and the Genesee Farmer every day and week, *just as far as he dares to.* Such treason to freedom is none the less pernicious in its effect for being cowardly and insidious in manner, none the less detestable and outrageous for being ridiculously weak. We will not say that the traitor should be scourged out of a community of freemen. But we do wish he might be, gently, "carried back to"—the place where he belongs. He has been trying to sell himself South for sometime past. What a pity that he could not find "pius Joseph" green enough to buy him, and put him under a course of his favorite partriarchal discipline! We fancy it would be a discipline suitable to his case. For Solomon says, "A whip for the horse, a bridle for the ass, and a rod for the fool's back." And a wiser

[man] (in his own esteem) than Solomon has written "To lift him [the slave] into the sympathies of civilized life, there may be better things than the plow, the hoe, and the *whip*, but neither ancient art nor modern science has discovered them."[53]

This was strong language, even in the times of editorial feuding, and Lee could hardly afford to ignore it. But first he was interested in finding out who his detractor was, who was engaging in such personal abuse. Lee asked the "conductors" of the *Democrat* to give him the name of the person who had been making these abusive attacks, but the *Democrat* steadily refused. Apparently Lee was prepared to use more than words on him, for he condemned the paper for standing between him and the "chastisement" which the "miscreant" deserved, who was "a coward as well as calumniator." He was "an assassin of character," and Lee demanded that he "avow his name and give his residence."[54] Lee finally gave up and placed him beneath contempt.[55]

Unable to use his fists, Lee had to resort to words, which he had always intended to use as supplements to his fists. "By suppressing important portions of the article in the April number of the *Cultivator*, from which it pretends to copy, (the Democrat) misrepresents our language as whole and meanly attempts to make us responsible for sentiments which we have never expressed, and do not entertain."[56] In answer to this charge, the *Democrat* a few days later printed Lee's complete article and attacked him further, heading his piece "Last Words with Dr. Lee."[57]

In his answer, Lee explained: "Our 'principles' are no more Southern than Northern—no more calculated 'to vindicate slave labor' than hireling labor. . . . The *truth* is our aim—nothing more, nothing less." He had never written a word in a Northern paper that he would not be willing to publish in his *Southern Cultivator*, nor had he written a word in the *Southern Cultivator* "against which any Northern man can reasonably take exception on moral, political or sectional grounds." The antislavery men in the North "would at once annihilate every institution in the world that differs a shade from their own. There is no tyranny like that of a full blooded fanatic."[58]

Lee's expression "the plow, the hoe, and the whip" had greatly enraged the *Democrat*, which abused him soundly for advocating such a barbaric method of controlling slavery. Lee replied that the *Democrat* was using it entirely out of context. The *Democrat* was little less upset by what it termed the insult of calling Northern workmen "hirelings." Lee answered that Noah Webster's dictionary implied no derogatory meaning in its definition of "hireling," and that the Bible implied no insult when it referred to the laborer being worthy of his hire.[59] And as for the treatment that Southern slaves got in comparison with the opportunities for work that the free Negroes in the North got, he cited a little incident in which he was involved on the streets of Rochester: Recently "an able-bodied colored man stopped . . . [us] in the street and begged money to buy bread for his family. His story was that although able and willing to work, he could not find employment at such wages as would support himself and family. . . . From his language and civility, we knew at once that he was reared at the south, and he owned that he was a runaway slave. He had his freedom, but no bread; he had able hands, which God gave him, and was willing to be any man's temporary slave, who would feed and clothe him, but he could find no master." Lee gave him a few coins and passed on.[60] As a general observation on the opportunities the Negroes in the North had, Lee said, "To exchange the legal right to work and live comfortably in Virginia for the legal right to starve or beg in New York or Canada is a step back in civilization."[61]

Lee did not stop to repel every attack made by a Northern newspaper, but he said this further about the *Democrat*, including the *Rochester Union* which also had been disagreeable: "Such political quarters as the *Democrat* and *Union* have but one kind of medicine for all the ills that afflict the body politic, to denounce the people of fourteen states as 'our enemies.' "[62] Another Rochester newspaper, the *Union & Advertiser*, which did not like Lee, especially for his politics, charged him with trying to break up the Union because he had kindly sentiments toward the South. He said that he had borne these attacks for sometime, "but when an unprincipled libeller garbles ones language, puts the vilest sentiments of one trying the overthrow of the government, into his mouth, and garnished the calumny with many original falsehoods, the wrong is too outrageous to be borne in silence."[63]

It was time for the South to answer pointedly Northern criticism of its institutions and ways of living, for a supine attitude was "fast making the planting states the feeble, dependent colonies of the ever-aggrandizing, over-shadowing north."[64] In the heat of the days following Lincoln's election in 1860, Lee even went so far in his zeal of upholding slavery to say that in the course of time slavery would extend to the North and throughout the Union,[65] something that George Fitzhugh, the Southern extremist, had been predicting.

Getting somewhat out on the edge of the slavery controversy, Lee brought into his argument the matter of universal education, one of his favorite topics otherwise. One of the best defenses the South had against the onslaughts of the abolitionists was to educate the masses and, thereby, wash out of their minds the incipient feeling of inequalities that set them apart from the planter aristocracy. It would scotch any attempts to organize a free labor party among nonslaveholders. Furthermore, an educated working class, not to include slaves, was much more productive than the ignorant, illiterate one. Relating the subject to what he saw happen in Rochester after the establishment of common schools there, Lee reported that the wealth of the city greatly increased. The same was true in Scotland and in New England, but in Georgia there were 40,000 adults who could neither read nor write, and if this continued, he predicted disaster for the state. "If the natural endowments of every child were properly cultivated, nine-tenths of the crimes, diseases, vices and follies of mankind would disappear forever."[66]

"Man was not designed," Lee said, "to pass through life a mere animal machine—a living thing to toil with its muscles, eat, propagate, and rot. . . . It is the legitimate purpose of a good education to cultivate the Man as well as the Earth, out of which he was formed."[67] For twenty years, he said, he had tried to create a public sentiment for Congress to appropriate the whole public domain "to the sacred cause of public education." And in this he included aid to agricultural schools, and he predicted that the "careful study of soils, of cultivated plants, of domesticated animals, and rural affairs in general, will some day be as highly esteemed in this country, as the study of Greek and Latin now is."[68] "A good scientific agricultural education" would one day "be placed within the reach of every poor man's son."[69]

Farmers could learn, apply scientific findings, and get out of their old ways. It was an "insult to say that farmers could do nothing but repeat the action of young robins, keep their eyes closed, open their mouths, and swallow whatever is dropped into them under the name of science. . . . Farmers naturally look for the fruits of all scientific investigations that claim affinity with their profession; and unless good fruits are palpable and visible, agriculturists are apt to treat the assumptions and theories of school-men as idle speculations."[70]

Educated farmers would not be on the move westward as so many of them were at that time. They would take pride in their homes and make them a permanent abode. "A well kept garden and orchard, next to wife and children," Lee believed, would give "home its charm. To him who loves his garden, or orchard, or farm, every plant—every tree—is a friend with whom he communes."[71] Even this beautification could extend to plantations and farms. Lee said that the best hedge he had ever seen in the United States was near Augusta, a mile long, made by a Mr. DeLaigle, who was a refugee from the San Domingo slave uprising. The hedge was made of "the Cherokee Rose," which, when in full bloom, presented "a magnificent floral spectacle" and filled the "atmosphere with delicious perfume."[72]

## ❧ X ❧

# The William Terrell Endowment

DANIEL LEE was a friendly man. After coming to Georgia in 1847 to edit the *Southern Cultivator*, he soon became acquainted with the principal agricultural leaders of the state. Among them was William Terrell of Sparta who was the outstanding Spartan of his time, but in the old set phrase of praise, he was generally referred to, when occasion warranted, as "the noblest Roman of them all." Lee was to have special reason to hold Terrell in grateful memory.

Terrell's obituaries in the Georgia newspapers stated that he had been born in Wilkes County, Georgia, but his sketch in the *Biographical Directory of the American Congress* gave his birthplace as Fairfax County, Virginia, in 1778. The former statement is most likely correct, for his grandparents moved to Wilkes County with a family including grown sons, one of whom became Terrell's father. William Terrell grew to young manhood in nearby Hancock County, where he spent the rest of his life. He studied medicine in the Medical Department of the University of Pennsylvania.[1] And although practicing his profession when opportunity allowed, he came to be best known as a wealthy planter with plantations in both Hancock and Washington counties, amassing a fortune of almost a quarter of a million dollars, which measured in the dollars of the 1970s would be a million dollars or more.[2] Being a highly public-spirited man, Terrell served in both the House and the Senate of the Georgia legislature and also two terms in the United States House of Representatives (1817–1821). He played a prominent part in the Southern Central Agricultural Society of Georgia. At his death he had a library which appraisers, who, in the custom of the day, had little respect for books, valued at $1,500. Also in his home were busts of Washington and Franklin. On his plantations were at least 200 slaves.[3]

Terrell died on July 4, 1855. A Savannah newspaper editor in noting his death said that Terrell was "well known, not only as an enlightened agriculturalist, but as one of the wealthiest and public spirited citizens of Georgia." Continuing, the editor paid him this tribute: "As a man of refined taste, high mental culture, and extensive and varied information, he had few, if any, superiors in Georgia, while the qualities of his heart formed in all respects a fitting counterpart to the high endowments of his head."[4] The Trustees of the University of Georgia in special appreciative remembrance of Terrell for his munificence to that institution resolved a suitable speaker should be chosen to deliver an address before the Senatus Academicus (the official body in charge of higher education in the state) on the life and public services of Terrell.[5]

Terrell's gift, which was to tie Lee to the University for almost a decade, was foreshadowed by a movement to promote the teaching of agriculture in the state which had been started before Lee came to Georgia, but in which he vigorously participated after his arrival. In 1846 the *Southern Cultivator* carried an editorial (presumably written by Editor James Camak, but possibly written by Lee) announcing a gift to Yale College by J. P. Norton of $5,000 to be used in establishing in that institution a professorship of agricultural chemistry, provided $20,000 should be raised to add to the endowment. At this time a bill was before the Georgia legislature providing for an appropriation of $2,500 annually to promote agricultural education in "state colleges" but as Lee remembered it, the money was to be used to "found and sustain an Agricultural Professorship in the State University."[6] As this was before Lee assumed the editorship of the *Southern Cultivator* and while Camak was still living, he must have been on one of his travels, which he so much liked, as he said, that he attended a meeting of the legislature. So, the *Southern Cultivator* editorial announcing the Yale gift and the failure of the bill in the Georgia legislature might well have been written by Lee, and especially so since the last sentence was typically Lee language: "The legislature met—wrangled their usual time about party politics, passed laws for private ends, or, perhaps, for the benefit of what they call the learned professions; but not the first syllable was uttered, so far as we know, in reference to enlightening, elevating and ennobling that profession whose great business it is to convert earth, air and water into bread, meat and clothing."[7]

The Yale professorship made a deep impression on Lee, for it was carrying out what he had been working for in New York for years, and in his *Genesee Farmer* he advised young men, who were able, to attend the Yale lectures.[8] Terrell was equally impressed

with what Yale was doing, and in 1851 he began thinking what
likewise he could do for the University of Georgia; but before
proceeding he tentatively offered Lee such a professorship in the
University, if arrangements could be worked out.[9] Nothing came
of this, but three years later, Terrell made a gift to the University,
which added to his fame at least for a time, before he fell into
oblivion, for both a professorship and a building on the campus
were named for him, as well as one of the counties in the state.

So, in 1854, Terrell proposed to the Trustees of the University
of Georgia a gift of $20,000, the interest from which should be used
to pay the salary of a professor whose duty should be to deliver a
course of lectures on "Agriculture as a Science; the practice and
improvement of different people; on Chemistry and Geology, so
far as they may be useful in Agriculture; on Manures, Analysis of
Soils and on Domestic Economy, particularly referring to the
Southern States." The lectures should be free. If the gift were
accepted, then Terrell asked the privilege of recommending for the
position "Dr. DANIEL LEE, who has spent twenty years of his life in
the study and practice of Agriculture, and who will bring to its
duties, all his skill and a zeal that ought to ensure success."

In explaining his reasons for making the gift, Terrell said that
agricultural education was being neglected all over the United
States, but especially in the South, where with "the advantage of
soil and climate" agriculture was far behind the Northern and
Eastern states. In the name of patriotism and national safety, he
told the Trustees, "The best form of government for a country
where a system of agriculture prevails that is constantly tending to
impoverish the soil, cannot long sustain a thrifty population, or be
able to defend itself. To avoid such a calamity, which there is reason to fear will be our condition at no very distant day, the people
of the Southern States must find the means of preserving their land
from destruction by bad tillage, which is so strikingly observable in
every part of the country." [10]

With great pride the Trustees accepted the gift, honored Terrell
by naming the chair the Terrell Professorship of Agriculture, and
appointed Lee to the position. The gift, the Trustees said, was "a
manifestation of enlightened public spirit, unprecedented in the
history of Georgia, and we believe we may safely add, in the Southern States. It is seed sown which will yield to the people of Georgia,
harvest after harvest in coming years." It would bring science "to
the help of agriculture, the great industrial pursuit of our people,

giving promise of reclamation of waste lands, fertility to exhausted soils, and a new impetus to southern enterprise."[11]

The Trustees appointed a special committee to prepare an announcement of this gift, in which the nature and importance of this endowment would be set forth. It said that the trustees would "attempt to confer no praise on that which is above all praise; but honored with the administration of such a trust, they cannot withhold their homage of the administration of an act of such pure and elevated patriotism—one which indissolubly connects the name of the donor with the great and commanding business of the State, and places him in the enviable position of a benefactor whom posterity will delight to honor."[12]

Agreeing completely with Terrell, who made the gift, and with Lee, who was to carry out its purposes, the Trustees' special committee noted that although Georgia was "still a young community, already a large portion of the soil is reduced to a state of absolute exhaustion." And, under conditions as they then existed, it was "notorious that the cultivated lands are becoming poorer"; if nothing was done to change the situation, "the time is not distant when the scarred and barren fields of the older counties will be the type of our whole territory; and a sterile soil and an impoverished people, be the characteristic of the State, now proudly arrogating to itself a leading position among its sister republics." It was not long since, said the committee, it was good planter philosophy to hold that agriculture was "better served by clearing a new field, than by renovating an old one."[13] In Europe the people had been working their fields since the time of the Romans, and by the use of scientific agriculture in some places they had more than doubled their yield within the past half century. Georgia had much to learn, and Lee in this new professorship would tell them how: "In draining, in irrigation, in deep and careful ploughing, in the preparation and application of manures, in the rotation of crops, in the use of labor-saving machinery, and in the selection of breeds and rearing and fattening of livestock, the agriculture of the State is as yet comparatively in its infancy."[14]

The committee was aware "of the prejudice which unfortunately exists in the minds of many practical men against science, and how much such a prepossession will impair the usefulness of the endowment," but it argued that scientific agriculture was not simply book learning but something based on practice, observation, and experimentation.[15]

As news of this donation spread, it created great interest and even excitement in Georgia and over the South. First of all, the students in the University looked forward to the time when they would begin listening to Lee's lectures. They termed the gift a "magnificent donation" and Lee "universally acknowledged to be a ripe scholar and a practical man,"[16] with a "well established and wide-spread reputation."[17] An Augusta newspaper editor placed Terrell in the company of that illustrious few Georgians who had made donations to education—John Milledge, who gave the land on which the University was founded, and Jesse Mercer, whose estate helped to bring Mercer University into existence. These men had done more for the state than all the politicians who "have or ever will demagogue it in Georgia."[18] The editor of the Savannah *Courier* said that "no other person in this great Republic has given for immediate use, to increase and diffuse rural knowledge, more than one-fourth the sum donated by the patriotic and distinguished founder of the first Professorship of Agriculture in the Southern States."[19] In a reference to the Norton gift of $5,000 to Yale College, through which Norton's son was awarded the professorship, the Savannah editor, in noting Lee's appointment, said that no "paternal or family tie" was attached to Terrell's gift.[20]

In Tennessee, Dr. F. H. Gordon, proprietor of Sugartree Farm, wrote Lee, "Your Agricultural Professorship may be regarded as the beginning of an era—a reformation, which will make a powerful impress upon the character, intelligence, wealth and future destiny of all the States."[21] In Mississippi, a newspaper editor said that Terrell "did more good for the people of Georgia when he established the Agricultural Professorship in the State University, than if he had established a dozen medical colleges, with as many law professorships." He said further that it was in the schools and colleges where "the *principles* of agriculture" should be taught. And he expressed the hope "that a new era is dawning in the history of cultivating the soil in the South. If other States and other colleges will do as the State University of the Empire State of the South has done, the business of farming will soon be as tempting to the ambition of talented and educated young men as the pursuits of law and medicine." In referring to those who ridiculed "book farming" and agricultural journal preachments, the editor could not refrain from being very direct in answering them: "Many go so far as to ridicule agricultural journals, and call them humbugs, so conceitedly wise are they in their profound ignorance. Had these men had training under some able agricultural chemist, like Dr.

Lee, of the Georgia University, and mingled with the dog Latin and worse than dog French which Alma Mater manufactured for them, a little knowledge of the constituent parts of soils and minerals, the laws of vegetation, the decomposition of matter, the influences of light, heat and moisture, they would have had very different ideas, not only of their business, but of themselves."[22]

Up to this time (1854), Lee was editor of the *Southern Cultivator* (Augusta, Georgia), editor and owner of the *Genesee Farmer* and coeditor and part owner of the *Rochester* (N.Y.) *American*, and agricultural expert and editor of the agricultural section of the Patent Office reports (Washington, D.C.)—with all of this geographical distribution of duties, Lee had been a commuter among these three cities (with his Washington position ending in 1853). Now, in accepting the Terrell Professorship of Agriculture, he divested himself of the editorships and ownerships of his Rochester publications and became irrevocably and permanently a resident of the South. In leaving the *Genesee Farmer* he said that he would "not be separated from his old friends, the readers of the GENESEE FARMER. We owe them a debt of gratitude which will only be cancelled when life's labor is past."[23] But thereafter Lee never contributed to that first love of his, no doubt because he fell into disfavor with many people of Rochester and the state because of his slavery views. He continued as editor of the *Southern Cultivator*, with his assistant Dennis Redmond now being added as a full editor. In noting Lee's appointment to the Terrell professorship, Redmond said, "No man in this country has labored longer or more faithfully for the cause of American agriculture than Dr. Lee, nor do we know of one better qualified to fill the honorable position to which he has been called."[24] In accepting his new position, Lee said that it was his "purpose to do all that may be reasonably expected to render it alike creditable" to the University, "useful to the public, and an honor to its founder."[25]

# ❧ XI ❧

# Professor of Agriculture
# at the University of Georgia

DOCTOR, EDITOR, PROFESSOR—Lee qualified for all these titles, though he had long since given up active medical practice; for the next eight years he would be a professor, and for the rest of his life he would be an active editor or an associate editor or a correspondent for various agricultural journals, and for a short time an agricultural editor of a literary journal.

The first announcement of his professorship in the University of Georgia stated that he would begin his lectures on January 15, 1855, but the time was later changed to the first Tuesday in March. His lectures were, of course, to be given "to the students of College" and also to "such other persons as may choose to attend"—all free of charge. He was not required to perform police duty over students (as was the case of all other members of the faculty), and his salary would be $1,200 annually, which was the income from the endowment of $20,000 at 6 percent, or more if the interest rates made it possible.[1] The other members of the faculty received $2,000, but Lee's salary was, in a sense, better than the others, for his University duties required only about half of his time. His lectures were to be given on Tuesdays and Thursdays; all seniors were required to attend and be examined each day on the subject of the previous lecture. Before ending each lecture Lee would announce the subject of the next one.[2]

Lee gave references to books in the library to be read, and, although he used one book rather constantly, it could hardly be called a textbook. He announced in his first lecture that he had been contemplating and working on a textbook intermittently for many years but that he had not yet got it in order for the printers. And on hearing of Lee's lectureship, a Tennessee printer begged "in behalf of the agricultural interest, that Dr. Lee shall set about at once and write a text book on agriculture adapted to schools. Such a book is needed in the University, and will meet with ready sale all over the country."[3]

Now, for the first time there appeared in a catalogue of a Southern university or college and probably of only one Northern college a section that indicated the inclusion of the study of agriculture in its curriculum. This advancement in higher education was suggested by the following course description: "The Terrell Professorship of Agriculture. Lectures will be delivered on the following subjects: 1st. Agriculture as a Science; 2d. The Practice and Improvements of different peoples; 3d. Chemistry and Geology, so far as they may be useful in Agriculture; 4th. Manures; 5th. Analysis of Soils; 6th. Domestic Economy, particularly referring to the Southern States. It will be the aim of the Lecturer to develop the true principles of Tillage, Husbandry and Plantation Economy, with a view to their general practice, and the attainment of profitable results. The Lectures are to be free to all who may wish to attend."[4]

The University and the town looked forward to Lee's first lecture. He realized that he had a delicate task in pitching the level of his lecture to a diverse audience of students and faculty, intellectuals and ordinary citizens, and the mechanics of the town and community. He labored successfully to be scientific enough not to deal in trivialities and yet to talk neither up or down to the audience and to make the lecture interesting enough for all to hear. He organized his lecture around a preliminary discussion of the subjects Terrell had mentioned in his letter of gift. In going into "the arts and science" of agriculture, he explained that in essence "an *art* is always something to be done; a science is always something to be *known*." He made plain how theory could be put into practice, but that after all, theory (or science) was developed out of practice.

Lee was, indeed, a stern believer in practicality in farming, for wherever he went and stayed long, he developed a farm and combined dirt farming with his scientific investigations and applications. In his lecture he emphasized the importance of analyzing soils and knowing what had to be replaced in soils as crops removed essential elements. He also emphasized the necessity of knowing something about geology—how the rocks made soil and how important the different kinds of rocks were. In discussing geology, he was at pains to say that there was no conflict between Genesis and Geology, for it was only the interpretation some people put in geology that would indicate a conflict. He made some homely remarks on domestic economy, by which he did not mean stinginess but getting the best out of resources at hand—how in heating a house much wood and labor might be wasted in improper construction of fireplaces. He noted in this lecture that many subjects, which he was merely suggesting now, would be taken up in the future. Paying a compliment to his adopted state, he said, "Georgia has every natural advantage, wealth and population enough, to be made into something like a paradise to live in. Its agricultural and horticultural resources have never been investigated with that degree of care, caution and patience which equally avoid exaggeration on the one hand, and a hasty underestimate on the other. Facts are sober realities; and they should be soberly considered by all who would understand their true meaning."

Lee ended his lecture with this hope and admonition: "By making a wise and successful use of the means of improvement which we already possess, we may reasonably hope that the Legislature, or some noble friend of agriculture, will give to the University of Georgia funds sufficient to purchase stock and equip a farm for experimental purposes, that Practice and Science may together develop and demonstrate the best system of Rural Economy for the Planting States."[5]

Enthusiasm for this lecture, and subsequent ones, was widespread. The senior class, as was largely the custom among the students in connection with important outside lectures, asked Lee for a copy of the lecture so that it might be sent to the agricultural journals of the state and the South for publication. Lee provided the copy which his own *Southern Cultivator* published[6] and which other journals reprinted. The *Georgia University Magazine*, the University student publication, said the students had "listened with much interest" to Lee, who had proved to be "well versed in Botany, Geology and Chemistry."[7] Over in Columbus, Georgia, the editor of the *Soil of the South* welcomed the lecture after having received a copy from Lee and said that it was a "satisfactory fulfilment of the designs of its munificent founder" and that Lee as an agricultural writer had "no superior in the country."[8]

President Alonzo Church of the University, who, in the salubrious Southern climate and even with a Southern wife, was never able to outgrow his New England puritanical reserve, said after listening to a few of Lee's lectures, "I was pleased with his lectures as far as I heard them and have little doubt that they will be interesting and useful to the young men and to others who may be disposed to attend them."[9]

In his lectures Lee wandered widely over the subjects of agriculture, rural economy, and their related fields. In his "Lecture on Labor" he took a historical and philosophical position, as well as a practical one. He made it quite evident that man was designed from the beginning to earn his living by the sweat of his brow; it was part of his very being to do so. A system that taught that man could live without work was an evil: "Persons calling themselves philanthropists not only 'rejudge the justice' of Heaven, but would fain re-create the universe to give their fellows a happier existence than this world affords." To his mind, nothing in nature was "clearer than the fact, that our daily wants of hunger, of nakedness, of sleep, and of the shelter from the extremes of heat and cold, are designed by Providence to make us preeminently *working* as well as *thinking* animals." Reverting to his slavery topic, of course, he did not have to defend slavery before a Southern audience, but only said that it provided that additional labor which the South had to have and that the so-called slave was in fact a "negro apprentice for life." He might have added here that it provided the social security which was denied the Northern free laborers.[10] His lecture entitled "Lime" was of such general interest that he gave it in the Presbyterian Church auditorium where is was easier for the townsmen and farmers and planters of the surrounding country to assemble.[11] Somewhat on the edge of an agricultural subject was his lecture "Hereditary Blood in Man and Other Mammalia."[12]

Always a humanitarian in his instincts, Lee hoped in his lectures not only to impart practical knowledge but also to uplift his students spiritually, to inspire them to do better work in their other subjects, and to make them better citizens in their chosen professions. Some of Lee's lectures were published in the *Southern Cultivator*, and a reader of that journal suggested that "it would be a great benefit" if all of them were published there. Admitting that there would be some value in that, Lee said that the reader would miss the personal contact, the questions and answers, and the readings required in the University library.[13] Without belittling the magnificent Terrell gift, another reader felt that these classroom lectures hardly provided what Georgia most needed. He said, "We very much doubt, if it be possible to unite at this day and at the same point a full course of classical and Agricultural education."[14]

Martin W. Phillips, a well-known Mississippi agricultural editor, much impressed by the value of Lee's University of Georgia lectures, thought that Lee should "deliver lectures through the length and breadth of the State, at every county site at least, and also at any point where an Agricultural Society is in being." The legislature should make an appropriation for all his expenses, including "chemical apparatus," and pay him "a fair salary" in addition to his income from the Terrell foundation. Phillips suggested also that Lee should be sent to Europe to see what was going on there along agricultural lines and then report his findings. Lee agreed with all these suggestions and especially in carrying his message to the people over the state, for "Explanations may be given where one talks to face, and thus remove objections and difficulties, which in writing for the press is comparatively impracticable." Writing in August of 1857, he had said he would visit "many counties before the close of the present year, and appeal to the people in behalf of their domestic policy which will make, if adopted, the South the garden of America."[15] In addition to carrying his lectures to the people over the state as opportunity afforded, Lee in his spare time took on other assignments. In 1860, he made "an agricultural survey of large tracts of land" in Florida, one of them being a 9,000-acre tract on the St. Johns River owned by a New York investor.[16]

Lecturing to students and people in general was only half the problem, if even that, in promoting agriculture and rural economy. To a person with Lee's zeal for experimentation and scientific research, he could be satisfied with nothing that did not include them. For thirty years, he said, he had urged the wisdom "of making experiments to increase the agricultural knowledge and prosperity of the country."[17] Science and practice rarely came "near enough to speak to each other. . . . In a word, theory is valueless beyond what experience fully sustains in practice."[18] "Agriculture is not an *exact* science; and its most important truths are reached by experiments."[19] Lee could hardly talk about agriculture without making such expressions.

So when Lee accepted the Terrell professorship, he was told by the Trustees of the University that his lectures should embrace the science of agriculture "and as much of the practice as can be taught in that way"; but the "limited means of an individual endowment, however magnificent, precludes the practical application of principles to the workshop or the field; but fortunately they are sufficient to supply the most pressing wants of the State, which is a general knowledge of the great principles in the science of agriculture."[20] After Lee had got started, a committee of the Trustees recommended that everything possible be done to promote his work—

"A suitable Lecture Room, Laboratory, and Museum are necessary if not indispensable."[21]

Terrell was as much in favor of experiments as was Lee, and in a letter to Lee he said, "The college ought to have and own all needful scientific apparatus, if it is to do anything useful; and without the means of illustration, you might as well write books on science."[22] Seeing that the University was unable immediately to supply laboratory materials, Terrell sent Lee a check of $500 to make such purchases.[23] At the beginning of his lectureship, Lee announced that a museum must be built up and a library of books "of strictly professional character" would have to be developed. A museum was indispensable for it would show the actual results obtained by the application of agricultural science: "Talking with no visible demonstrations, it is not the most approved plan of teaching science." What could be done in the practical part of farming would depend on public support and that of the legislature. Lee suggested that possibly "some one or more may give the funds required to purchase and stock a farm for experimental purposes."[24] He observed, "No mere verbal instruction, no chemical manipulations within the four walls of a laboratory, will satisfy plain, outdoor farmers."[25]

As for adding agricultural books to the library, Lee said in a letter to three of his Augusta friends that he needed at least $50 a year, with $500 immediately. His salary did not permit him to go to this personal expense, as he did need his expenses paid while in his spare time he worked "to demonstrate, in a practical way, by field-culture, stock husbandry and farm economy, the true value of *applied* science."[26] Lee illustrated his lectures with as many "chemical demonstrations" as his equipment made possible. For example, he burned cotton before the students to show how much air there was in it, to demonstrate that a bale of 500 pounds contained 495 pounds of air.[27]

In making his gift to the University, Terrell fully expected that the legislature could be induced to add to it, for he knew well that the purpose of his gift could not be carried out with the income from his endowment which could not provide for a laboratory, a museum, an agricultural library, and an experimental farm.[28] As Lee put it, Terrell "wished to kindle something like a vestal fire in the State of Georgia that should forever keep alive, at one point at least, an earnest devotion to the improvement of southern agriculture."[29] Immediately the cry went up for the legislature to act. The University Trustees backed by the faculty, petitioned that body in its session of 1855–1856 to add $20,000 to the Terrell Endowment and thus make it possible to give Lee full time on the faculty with a salary equal to the other members.[30]

Gov. Herschel V. Johnson, in his message to the General Assembly, recommended an addition to the Terrell Endowment and overenthusiastically assured the state that the appeal would "not be in vain." He said that there was a special need for the proper equipment of the professorship, for Professor Lee might "delight his class with the theory of Agriculture, but he must have the means of illustration and experiment to unfold its relations to, and dependence upon" chemistry, botany, mineralogy, and the other sciences connected with agriculture.[31] To promote the good cause, Lee appeared before a night session of the General Assembly and enlivened his address with various simple experiments.[32] The Committee on Public Education and Free Schools reported a recommendation to add $10,000 to the Terrell Endowment, but the movement died stillborn.[33] Disappointed, Lee later wrote, "For a great agricultural community, like Georgia, to pursue a policy that forces its most active and useful citizens and capital to emigrate from the wanton impoverishment of its arable lands, is simply to make the State commit suicide."[34] He did not give up hope of the General Assembly's doing something, but he began to think of individual gifts as a more likely source of funds that would make possible the realization of Terrell's vision—and Lee's too.[35]

True enough, as part of what Terrell and Lee had in mind along experimental lines, the University announced from year to year that it had "a valuable and very complete Philosophical and Chemical apparatus, with which the professor is enabled to give illustrations of all important subjects in Experimental Philosophy" and that it had also "an extensive Minerologocal [sic] and Geological Cabinet."[36] But at least the chemical apparatus could little serve Lee's needs, since according to the announcement of the Terrell professorship, Lee was to lecture on "Chemistry and Geology, so far as they may be useful in Agriculture."

So when Lee entered upon his professorship, he was much concerned about experimental apparatus, a museum, a library, and other aids useful in teaching agriculture. Although he was included in the regular faculty list, he was not on the same basis as the other members of the faculty, for his salary of $1,200 was received from a special fund, only half of his time was required for his lectures, he was not required to attend faculty meetings, and he was free from police power over the students. From the beginning, both the

University Trustees and the faculty felt that Lee's special status should be changed to make him a regular member of the faculty and let him give full-time service and receive the $2,000 salary which others got. But the trouble was, where would the extra $800 come from? The Terrell Endowment provided only $1,200 and the University ran on a very tight budget, made up of student fees and the $8,000 annual appropriation of the legislature, which was not a gift at all but interest on a fund that the state had years before taken from the University. The Trustees thought that the only way out was to petition the General Assembly again for an addition to the Terrell Endowment—this time $15,000 in 7 percent state bonds.[37] This appeal failed too.

In a general reorganization of the faculty in 1856–1857, a re-hiring and regrouping of the faculty made possible funds sufficient to bring Lee's salary up to the $2,000, but this change in his status was not made until 1859. Now the subject of chemistry (not simply agricultural chemistry) was added to his teaching load, and his title now became "Terrell Professor of Agriculture, and Professor of Chemistry."[38] Chemistry had been previously taught by other professors and was scheduled for juniors in the third term and for seniors in the first term. This arrangement greatly restricted Lee's freedom to look after his agricultural interests, for the first term (the University year was divided into three terms) began in August and extended to November and the third term began in April and extended to August. Vacation began in November and extended to January. Thus, Lee's commitment to teaching chemistry took up the time when farming was being carried on, that is, from April to November. And in addition, he would find it very difficult to continue his editorial work on the *Southern Field and Fireside*, which he had taken on in 1859 when he gave up his editorship of the *Southern Cultivator*. Now he would be required to police the students, which was so time-consuming and distasteful to some of the faculty that it had led to a general disruption in 1855–1856, resulting in the loss of some of its most distinguished members.[39]

Lee said that this new dispensation was in complete violation of the terms of the Terrell gift: "Wise and good men who make laws, will not break them for trivial purposes."[40] The University had not spent a dollar to provide a lecture room for him, Lee said, nor had it bought an agricultural book for the library nor spent a penny to promote his experiments. But the patronage that had come to the *Southern Field and Fireside* since he had joined that publication had made it possible for him "to commence a series of experiments

with numerous grasses, and sheep husbandry" with promising results. Also in furtherance of this work he had contracted with an
Athens livery stable to furnish him the manure from the stalls of
sixty horses for a period of five years. It now seemed that he would
have to give up his professorship or his editorial work on the *Southern Field and Fireside*.[41]

Lee won his point; chemistry was divorced from his professorship and his salary was reduced to its original $1,200. But otherwise
he was to be considered a regular member of the faculty with the
right to attend faculty meetings, to vote on all measures, and to
have equal authority over student behavior—but he was not required to police students. And in answer to his requests, the Trustees agreed to furnish a room for his agricultural museum and to
refer to a faculty committee his desire to make "a contemplated
visit to foreign agricultural colleges," provided that the University
would not pay his expenses but that any documents and information he obtained be given to the University—which, no doubt,
pleased Lee, for he would use this information in his lectures;
any material things would go into the University agricultural
museum.[42]

Not only were the strains and stresses in the nation becoming
more taut, which would soon break out in war, but also there was
no dead calm within the University Trustees, and especially on the
subject of Lee. In June, 1861, Thomas R. R. Cobb introduced a
resolution in the Trustee meeting calling for a discontinuance of
the Terrell professorship at the "end of present term . . . as at present organized" so that the income from the endowment could be
allowed to accumulate "during the present trouble of the country."
This could hardly be considered otherwise than a thrust at Lee,
and it "elicited much discussion," which resulted in passing a mo-
tion to lay the matter on the table. Supporting Cobb in his fight
against Lee was Cobb's brother Howell, who had recently been
Secretary of the Treasury in President Buchanan's cabinet; former
governor Wilson Lumpkin; and other prestigious Trustees from
Athens.[43]

Lee's comment on the move to oust him was that the Terrell professorship was now "the living germ of a noble idea, originating in
the cultivated common sense of its founder." Destroy this "intellectual germ by abolishing the Chair . . . [and you] affirm that in
southern agriculture, knowledge is not power, nor desirable. . . .
It is absurd to expect to dig sound principles out of the ground like
a basket of potatoes." Scientific principles in agriculture must be

taught and found out by experimentation. Schools were set up to teach law, medicine, and other professions, but why not agriculture, Lee wanted to know.[44]

Even in the face of this hostility of two of the principal architects of the Southern Confederacy (Howell Cobb, president of the Convention and Provisional Confederate Congress, and Thomas R. R. Cobb, principal framer of the Confederate Constitution), who five months previously had played their part in setting up this new government, Lee, a synthetic Southerner, was to make a greater personal financial sacrifice than any other member of the University faculty. He proposed to remit one-half of his salary in the interest of the University, Georgia, and the Confederacy. The Trustees accepted this substantial voluntary contribution and passed a resolution cutting the salaries of the other members of the faculty 20 percent.[45] A year later (July 4, 1862), those who had wanted to get rid of Lee succeeded. The Trustees resolved that after January 1, 1863, the Terrell professorship "be vacated, until it shall be the pleasure of the Board to fill it," and that in the meantime the income from the Terrell Endowment be kept separate and added to the principal.[46] By the following summer, the University, both faculty and students, had become so involved in the war by their entering the fighting forces that the institution found it impossible to open in the fall. It did not convene again until 1866.[47] Lee, who had held either two or three positions during the past dozen years, now found himself without even one, for the *Southern Field and Fireside* suspended in November, 1862, and when it resumed in 1863, Lee was elsewhere.[48] After the war, Lee said that he left Georgia "when it became unpleasant for a Northern man to hold a Professorship" there.[49] So ingrained in the consciousness and subconscious of some Southerners that all Northerners were suspect and not to be trusted that it naturally followed that Lee could never qualify with them as a Southerner.

But one factor in Lee's life and services in Athens and at the University of Georgia has yet to be noted. In line with his passion for experimentation and his determination to enter into it wherever he might be for a few years, aided or unaided, he must have a farm of his own. So, by the year after he began his lectures in the University, he had sufficiently spied out the country surrounding Athens to have hit upon a historic but worn-out farm of about 600 acres (to be exact, 596 "more or less"), to the west on the Jefferson Road (the highway to the town of Jefferson where Dr. Crawford Long had made his epochal use of sulfuric ether as an anesthetic in

an operation). This farm had much earlier belonged to Henry
Jackson, a brother of the famous Revolutionary hero and governor
of the state, who had come from England in his youth and who in
1811 became a professor in the University. Soon thereafter, on
leave, he accompanied as secretary of legation William H. Crawford, who went to Paris as minister to France. On his return, Jackson continued in the University and on his retirement lived on this
farm, which he called "Halscot"; he died there in 1840. His son
Henry R. Jackson, later to become as famous as any of the Jackson
tribe, sold this farm to Pleasant Stovall; it was from Stovall that
Lee bought it for $2,400—$4 an acre.[50]

Here was an excellent opportunity for Lee to apply his scientific
skill in experimenting with soil restoration, growing grasses, engaging in dairying and other livestock business, raising various crops,
developing orchards and timber resources, and otherwise in making
the most use of worn-out and unimproved land. In 1860, he could
report to the United States census taker that he had improved 200
acres, but the remaining 400 acres were unimproved old fields and
timberland. He gave the value of his real estate at $5,000 and his
personal estate at $3,000.[51] His farming operations were modest
(which does not necessarily indicate any inadequacies in his experiments) as is shown by his answers to the questions asked on the
census form. He valued his farming implements and machinery at
$55; he had one horse, two milch cows, two work oxen, four other
cattle, and ten hogs. He valued all this livestock at $207. At this
time (August 25, 1860), he had on hand 400 bushels of corn, 100
bushels of peas and beans, 20 bushels of Irish potatoes, 125 pounds
of butter, and 4 tons of hay. In his smokehouse he had $76 worth
of slaughtered animals.[52]

He lived on his farm, a walking distance of "between four and
five miles from the Athens depot,"[53] or a little more than three
miles from the University. He lived in the plantation home which
had been occupied by Henry Jackson. With his instincts, naturally
he loved the country, but he had high praise for the town of Athens
and its inhabitants: "No other town in the South can exhibit so
many beautiful residences according to population, or a more refined society."[54]

At this time his family consisted of himself, his wife Sarah V.
Lee, and their son Henry Lee, who was four years old. Sarah was
twenty-two years younger than her husband (who was fifty-two
years old, according to the United States census); she had been born
in Washington, D. C., where Lee had undoubtedly met her when

he was in the Patent Office. They were married about 1855. This was Lee's second marriage; by his first wife he had several children, some of them grown by this time; one, Daniel Fitch Lee, being in charge of Lee's District of Columbia farm.[55] His first wife, Sabrina Lee, two years older than he was, died at her sister's home in Cayuga County, New York, in June, 1854, sick with dropsy for more than a year. She was sincerely mourned by Lee, who said that "no mother and wife was ever more obedient to every duty which GOD in His providence assigns to the most sacred family ties."[56]

True to Lee's support of the institution of slavery, he had a family of slaves on his farm, with the head helping him in sowing, planting, and reaping.[57] Lee's principal interest was experimenting with grass on his cleared acres. He had collected in this country and imported from abroad sixty different grasses.[58] He set out twenty acres of orchard grass and timothy and had his slave scatter guano over it. However, the summer proved hot and dry, and on turning his cattle and sheep on it, they nearly gnawed it out of the ground; still he had a good stand of it.[59] The editor of the Athens *Southern Watchman*, on venturing out to take a look at Lee's operations, said that if Lee's attempt to introduce these grasses to Georgia failed, then there was "no need for any body else to try it." If it was a success, and he thought that it would be, then it would revolutionize the whole farming picture in the state. Instead of gullies, red hillsides, and brown broom sedge, "we will have green pastures, handsome meadows and profitable flocks and herds—nice rich yellow butter and fat beef and mutton!"[60] What the editor described did happen a hundred years later. Ever with experimenting in mind, Lee turned his cattle on a 75-acre pasture of broom sedge, taking them off a pasture of orchard grass, and he found that the milk and butter dropped off more than one-half. Restored to the orchard grass pasture, they regained the loss.[61]

Interested in bluegrass as well as in orchard grass and timothy, Lee found that this grass had been growing on the University campus for thirty years, and he located patches of it in various places over Clarke County. In fact, he found some of it already on his land and on lands adjoining his, around the historic Gum Spring, "a bold and famous spring." This grass had got started through the hay which wagonmen from North Carolina and Tennessee camping there had brought along.[62]

Experimenting further with grasses, Lee had Oscar Bailey, whom he had secured to help him, sow bluegrass, orchard grass, timothy, and clover in a twelve-acre field of standing corn to shade the tender plants as they came up. He said that he had much success here and with other grasses at other times; but Bailey attempted to play down Lee and say that he himself was the one to whom praise should go. Lee successfully disposed of Bailey's claims, and added the information that he had known the Bailey family in New York before they had moved to Virginia and that Oscar had wandered on down to Georgia and was now attempting to tell Lee how to raise grasses.[63] Lee turned his little flock of sheep into his broom sedge pasture where they could graze on that tough grass and briars, and he found that they did very well; but they did much better, of course, on his developed pasture of grasses. It made him dream of the day when Georgia and much of the South would add grass to their rural economy.[64]

In carrying out his farming activities, Lee found time to develop an orchard and to utilize some of his timberlands. He set out 400 shockley apple trees "among others," which cost him ten cents apiece. Ten years later, after he had moved away, he learned that the trees were bearing and that the new owner was selling the fruit at $2 a bushel.[65] On one part of his farm he planted a few red mulberry trees, which he greatly prized for their fruit, an excellent feed for swine; the trunks made stout ship timber and fence posts "about as durable as yellow locusts."[66] And he could not refrain from boasting of getting forty cords of wood from one acre of his forest land, for which he had paid $4 an acre.[67]

Whether he remembered his professorship at the University of Georgia with any nostalgia, he could not forget the pleasures of his Halscot plantation, and when, in 1867, he passed by this place, he remarked that he "would rejoice to have a large stock and dairy farm near Athens, where our Youth might learn the arts of dairy and sheep husbandry, domestic economy, and to practice agriculture as a science."[68]

Lee had given eight years of his life as Terrell Professor of Agriculture in the University of Georgia, and he could not forget that. Hence, his interest remained and he kept track of what was being done with that position when the war had ended and the University had reopened. Whether he immediately learned these details is not known, for the catalogue for 1865–1866 left the professorship blank, though it repeated the page describing the professorship. In 1867–1868, the University began again using the Terrell Endowment income, adding it to the salary of William L. Jones, who was listed as "Professor of Natural Science and Agriculture," but it did not attach Terrell's name to the professorship, except that in a general description of the School of Agriculture it named Jones as "Terrell Professor."

This situation greatly displeased C. W. Howard the editor of *Plantation*, an agricultural journal published in Atlanta. He said that it was a misuse of the Terrell fund, since Jones was required to teach chemistry, mineralogy, botany, zoology, and geology, any one of which was enough for a professor to handle, and he gave only seven lectures on agriculture throughout the year.[69] Jones defended himself by saying that these lectures and work in other subjects related much to agriculture, and that he gave from ten to twelve lectures a year specifically on agriculture. Furthermore, he insisted that no competent person could be found who would teach agriculture for $1,200 a year.[70]

In 1872, the University of Georgia established the Georgia State College of Agriculture and Mechanic Arts from the sale of the public land received by the state under the Morrill Land Grant Act. The Terrell fund was attached to this college, since it was a division of the University, and the professor of agriculture in this college was termed "Terrell Professor." On the establishment of this new college, Lee felt that now was the time to fulfill his dream of associating practical agriculture with the Terrell lectures, and for this purpose he hoped that the citizens of Athens would give to the college 100 or 200 acres of land "on which five hundred boys can dig their grub out of the ground, while filling their heads with most useful knowledge."[71] Lee did not live to see the citizens of Athens and others, more than a quarter century later, add to a small experimental farm nearly a thousand acres to be attached to a new college of agriculture.

In 1872, Henry Clay White, primarily a chemist, came to the University, and two years later there was added to his position the income from the Terrell Endowment and the title "Professor of Chemistry, Mineralogy, Geology, and Terrell Professor of Agriculture"—this despite the fact that William Montague Browne a few years later was made "Professor of Agriculture and Horticulture, Natural History, History and Political Science," but was never put on the Terrell fund. White continued as the Terrell professor until his retirement more than a third of a century later. Lee, still keeping track of his old University, observed in 1876 in a disapproving spirit that the income from the Terrell Endowment had been added to the chair of chemistry and that the occupant had "been required to deliver a few public lectures on agriculture."[72] With the establishment of the College of Agriculture, the income of the

Terrell Endowment was added to the salary of the "Professor of Agronomy." Fortunately, the endowment was not lost in the catastrophe of the Civil War, and in the 1960s the fund was invested in state and federal securities. Thus, William Terrell's name through war and peace became fixed to a University of Georgia professorship; this was part of what Daniel Lee could have wished in recognition of the man whom he always delighted to honor.

# ❧ XII ❧

# In Time of War—The Confederacy

IN MAY of 1859, James Gardner founded the *Southern Field and Fireside* in Augusta and published it with one interruption into 1864 when it moved to Raleigh, North Carolina. He offered it to Georgia and the South as a protest against Northern editors and publishers who looked upon the South as intellectually barren, and only in a condescending air did they publish any Southern works. It would, therefore, be an outlet for Southern writers. As Gardner said, it was a journal for those (as well as for others) "who have shrunk from becoming mendicants for the favor of Northern critics and Northern publishing houses, and have wholly avoided literary pursuits, except as a pastime."[1] Although regarded mostly as a literary journal, it had two other departments: agriculture and horticulture.

In looking for an editor to conduct the agricultural department, Gardner did not have far to go in finding Daniel Lee. As Terrell Professor of Agriculture, Lee lived in Athens, but he was in Augusta frequently as editor of the *Southern Cultivator*. He was not hard to entice away from the *Cultivator*, for his salary there had dwindled until it was very little. So Lee gave up the *Cultivator* (but not his professorship) to go with the *Field and Fireside*. In announcing Lee's connection with his journal, Gardner said that Lee was "the distinguished Professor of Agriculture in the University of Georgia—editor for many years past of the *Southern Cultivator*, and a leading contributor to many Northern agricultural journals of the highest reputation."[2] Lee explained in his salutatory that he was leaving a monthly journal for a weekly one, where he would have a better opportunity to promote agricultural progress in the South. "One who has reached the down hill side of life," he said, "and somehow feels that he has yet much to do for the advancement of agriculture, may well seek increased facilities for the accomplishment of the work before him. It consists mainly in separating truth from error; and in making the former the common property of all."[3]

Secession and war were just around the corner when the *Southern Field and Fireside* got started. It fell victim to the war in 1864, but before going out, it had suspended in November, 1862, to the first of the following year. With the November 15 issue Lee's connection with the journal ended.[4]

In the meanwhile the coming of the war had put Lee to the test. Was he truly a Southerner, through thick and thin, or a New Yorker in the South parading as a Southerner for pelf and profit? Lee was not long in showing that his Southernism was based on principles that he had brought South with him and not a gloss that would wear off. And his Southernism was harder to maintain and defend than that which came from having been born and having lived in the South.

From his first residence in the South, Lee had had to suffer attacks on his being a Northerner and to defend his fixed principles, especially on slavery, which he had held all his life. From the very beginning extreme Southerners who were narrowly suspicious of everything Northern began saying that he was not to the "manner born" and asking what he knew about Southern agriculture and its needs. Despite Lee's all-out defense of slavery and his advocacy of reopening the foreign slave trade, some extremists began calling him an abolitionist. To this, Lee replied, "If fifteen years of unremitting labor with pen and often with tongue, in opposition to abolitionism and abolitionists make one obnoxious to the charge brought against the Editor of the *Southern Cultivator*, then he is guilty; otherwise the aspersion is as false as it is calumnious."[5]

Lee was at some pains to defend himself, and he had many broad-minded, intelligent Southerners to help him. Although not indiscriminately attacking his native North in defending his Southernism, he could say that the "sectional spirit of the North is nothing but an empty bladder, blown to its utmost tension, which a pin might puncture greatly to the relief of the patient,"[6] implying that the spirit might be owing to pecuniary purposes only. He could also say to Southerners, "It is not possible for him to change the fact that he was born in another State."[7] In a disagreement over certain chemical items with Dr. Joseph Jones, who was making a chemical report to the state, Lee was attacked by Jones with the innuendo that he was only an import from the North and did not know too much, despite the fact that Lee in his original criticism had paid a high compliment to Jones. Lee later said that Jones might have had higher compliments paid him, "but none that were more appreciative and sincere." But Lee did say of Jones's attack, "My real offense, however is not in what I said about lime salts, but on having 'originated in the northern section.' "[8]

Coming to Lee's defense a correspondent of the *Southern Cultivator* wrote, "I think more of you than any Northern man I ever saw";[9] and a South Carolinian said that "though my prejudices are very strong against Northern men, I am very willing to receive instruction in Agriculture from a Yankee or anybody that I feel is so well able to give it as I acknowledge you are." In answer to this compliment, Lee insisted that he was not a Yankee, "for the epithet (and we do not regard it as one of reproach) belongs, legitimately, to persons born in New England."[10] When the war broke out, there were others who acknowledged that Lee's teaching in agriculture and its diversification had made Georgia the strongest state in the Confederacy.[11]

How did the North regard this errant son who had gone native in the South? There were two lines of attack—one by contemptuous silence and the other by bitter denunciation of this "turncoat." Most of the papers that noticed him joined the hue and cry against him; but where one would expect some understanding of the man and certainly some mention of his name even if in a most contemptuous spirit, there was studied silence. For instance, after Lee gave up the *Genesee Farmer* he was scarcely mentioned, and during the war his name never appeared in its columns—even to be castigated. When the paper finally went out of existence in 1865 and a short historical account of the paper was naturally called for and given, the account managed to avoid mentioning his name, even though for years Lee had actually been "Mr. Genesee Farmer."

It seems hardly necessary to say that when secession was brewing and after war came that Lee was unswerving in his support of the South. He attuned his agricultural preachments to the support of the new nation: how agriculture could best promote the cause. Sometimes Lee almost forsook agriculture to pour out his condemnation of the North and its vandal armies invading the South.

Like many other good Southerners, Lee hoped that there could be found some way to settle disputes with the North without withdrawing from the Union. He toyed even with the chimerical idea that the North and the South without physically separating might in a sort of federation each rule itself—the South would take no part in sending up congressmen or electing the president, but in a way which he did not follow through on, he would have the South run its own affairs. In this way, he thought, the antislavery fanatics in the North could ease their consciences by knowing that their division of the government had no part in any way with protecting slavery.[12]

Slavery, after all, was the main bone of contention, and although President Lincoln wanted to destroy three billion dollars worth of Southern property, the North should realize that slavery was an absolute necessity both to the North and to the world at large. "The civilized world has not and cannot deny itself the great and obvious benefits to be had through the instrumentality of negro slavery." And, before the whole struggle over slavery would end, Lee predicted that some Northern states would see the advantage of adopting it.[13]

Rather than live under the hostile pressure of the North any longer, Lee said that separation, whether still in a federation or out, was "an unavoidable necessity."[14] Selecting the exact time was the only consideration, and if this were not conceded by the North, the South would be forced to resort to arms. But the South should be united in whatever it might do. If the South should be so, there would be no war, for the enemy would "concede to us the right to govern ourselves in our own way, and acknowledge our sovereignty and independence, without firing a gun against peaceful secession." At this juncture, Lee felt that separation need not be forever.[15] He could not understand why the North should not allow this solution. Let the South and North separate in peace as did Lot and Abraham in Biblical times, and let each live in peace "under its own vine and fig tree, with none to molest or make afraid."[16]

But Lee was to see otherwise, for soon the Northern forces crossed the Potomac and began an invasion of Virginia. Then Lee could say that "a more iniquitous war of invasion and subjugation was never perpetrated by human ambition and love of power." This was a "second John Brown raid," and to help repel it, the South should not allow a single bale of cotton to reach the North. The Lord would certainly punish the North for not letting the South go in peace, as He did when the Egyptians refused to let the children of Israel go.[17] Lee believed that the "madness of Northern journalists and statesmen" had brought on this catastrophe, which was "the most astonishing feature of the age." This invasion would destroy the whole economy of both sections, the South as well as the North, "for never was there a war of invasion before that was so certain to injure incalculably the invading party." By the time the first Battle of Manassas had been fought, Lee had come to the conclusion that the South could never again be associated with the North in the same government, for Southerners had lost all confidence in the Lincoln regime.[18]

Lee was for invading the North, seizing Washington and making it the capital of the Confederacy until a more central point could be selected.[19] And as the war grew hotter and Federal armies moved farther into the South, devastating the country, Lee came out for retaliation in kind. "Let the eight million freemen in the South," he said, "march an army into the free States large enough to destroy property, dollar for dollar, for all the injuries of whatever kind by Northern armies and ships to Southern property, rights, or interests; then our troublesome, meddlesome neighbors of the icy North will soon learn to mind their own business, and let the South alone."[20]

Now Lee pitched his agricultural program to wartime to serve the Confederacy; he warned that anyone who did not conform should be declared unpatriotic, if not a traitor. If no other good should come out of the war, the wartime economy of diversification and improved agriculture would be a peacetime blessing; it would have conquered the enemy and provided a protection against the North "for all time" if the South would "obey the dictates of common sense."[21] It was the duty of every farmer "to contribute to the support of our armies in the field, and to pitch his crops with that duty in view."[22]

Lee was in line with the general argument in the South that cotton would win the war, but that did not mean going all out for raising cotton, for withholding what was already on hand was the effective weapon—not raising more to entice Federal raiding parties to seize it.[23] Without stating it in so many words, Lee believed that more probably it would be food that would win the war, and especially if produced through scientific agriculture, which he had been preaching all his life. "Farm as we have done by growing two or three staples for export, and at the same time, create three acres of old fields for one in tillage, and our independence becomes shortly an absurdity."[24]

As professor of agriculture in the University, Lee believed that he had a fruitful field in the students who heard his lectures in which to plant the seeds for a better future. As they returned home or to the army most likely, they could carry the message of agricultural diversification. The young republic would "draw to its bosom the youth of the land, and its truest, warmest friends and surest support in every emergency."[25] Diversification included not only raising crops out of the ground but also the supporting of a live-stock industry, which was just as necessary in war, if not more so, than in peace. Lee thought that if the Confederates had had the cavalry forces required in any big battle, they could have captured or annihilated the whole Federal army and, for example, in the first Battle of Manassas have seized Washington.[26] For the saddler who needed to equip cavalry, the cattle industry should be en-larged to provide the hides for leather—to say nothing of the beef needed to feed the armies.

Still arguing for self-sufficiency for the Confederacy, Lee said, "We may show our independence by going barefoot, and riding horses without saddles, but it will be more effective and creditable if we learn to produce all the leather, shoes, boots, saddles and har-nesses that we need, and all the cattle required to supply the South with hides." Here it was easy and logical for him to make a pitch for his program of grass culture to make the pastures green. "Caval-ry horses, and gear for flying artillery, as well as harness for all private uses, suggest to thoughtful minds how silly it would be in any nation to depend entirely on other countries for the means of public defense."[27]

Green hides had to be made into leather. In the absence of com-mercial tanneries, farmers producing only a hide or two could band together and tan hides without difficulty. To explain how it was best done, Lee published specifications. He warned farmers against cutting down trees for firewood and other purposes at times when the bark could not be peeled off to be used in tanning activities. Both firewood and leather would be much cheaper when this fact was kept in mind.[28] In arguing for a livestock industry both in itself and as supplementary to farming, Lee would not let the young re-public forget the value of raising sheep, a subject that had long been on his heart and in his mind.[29] Like others at this time, Lee wanted the Confederacy to set up a department of agriculture, but he de-spaired that it would ever be done, for the politicians in the Con-federate Congress would never find time or develop the enthusiasm or understanding to do so.[30]

Agriculture and stock raising were not all. Iron and steel were a necessity both in war and in peace, and the South had the raw ma-terials for smelting pig iron, which with sufficient capital and proper skill could be processed into farm implements, tools, and weapons of war. While riding around on horseback in Clarke County, where his Athens home was located, Lee found beds of iron ore up the Oconee River, which flowed through Athens. He believed the ore could be mined and floated down to Athens after a few dams to aid navigation had been constructed. It could then be processed or loaded on railroad cars to be sent elsewhere.[31]

Mingling his voice in the outcry against speculators, who bought up whatever they could and ran the prices up to scarcity levels, Lee felt that the government, state or Confederate, should outlaw their activities and control prices.[32] As patriotic as any natural-born Southerner, and more so than many, Lee still could criticize the Confederate government for some of its acts or failures to act. Although a postage stamp was a small item in the budget of anyone, the ten-cent letter rate, increased from three cents under the old government, produced one of the first bad tastes in the mouths of the mass of Confederates and began the deterioration of morale, which was one of the main causes of the downfall of the Confederacy. Lee was one in that mass, and especially so because in his free distribution of seeds, he found no provision in the postal regulations that let him have a rate different from the letter rate. He cited as an instance where he had sent a bag of seeds to New Orleans (from Athens) by a special conveyance for $2. To have sent it by mail would have cost him $250 he estimated.[33]

The war wore on, and as Lee's duties as professor in the University came to an end before the year 1862 had become history, he increased interest in a white crystalline compound known chemically as sodium chloride (NaCl), but to most people it was common table salt. It was as necessary to all human and animal life as air and water, and no army could do without it any less than powder and shot, for armies were made up of men and horses and mules. It was common knowledge that all must have salt to live and that meats could not be preserved without salt.

Until the war started and the blockade was set up, the Southern states had got their salt from the North or abroad. Now, where could it be obtained? Lee set to work. In addition to evaporating sea water, the South soon began making arrangements to secure it from the salt springs at Saltville in southwest Virginia. Lee explored the possibilities of salt springs in Georgia, but found little except rumors. In midsummer, 1862, he set out for Saltville, Virginia, to see what could be done there. He went by rail on the Western and Atlantic Railroad through North Georgia and by other railroads on through Knoxville; after an initial visit to Saltville, he continued to Richmond—no doubt, hoping to see and talk to President Davis about the salt crisis. Apparently, he did not see the president, but on June 20 he addressed a letter to Davis, trying to impress on him the great necessity of doing something about the salt situation at Saltville where a monopoly that controlled the salt springs was making it difficult for individuals or even companies and state governments to deal with it. Lee advocated seizing the springs and having the government run them. He believed that solar evaporation could be used at Saltville, which would greatly reduce the cost of production of salt; he had seen it done in Syracuse, New York.[34]

Before Lee had to give up his ambitious plans of producing salt for Georgia at the Saltville springs, he said that he contracted for a thousand acres of woodland within four miles of the springs, from which to secure wood for firing the furnaces heating the evaporating pans. But when it came actually to setting up operations, he found it impossible to deal with the speculators who controlled the monopoly. One important result from his Saltville trip was that it caused him to leave Georgia and become a resident of Tennessee for the remainder of his life. In passing through East Tennessee, observant as he always was in his travels, he was greatly attracted by the countryside to the east of Knoxville along the Tennessee and Holston rivers. Indeed, he was much pleased with what he saw in Knoxville. So, before the end of the war, he bought property in that city and soon thereafter a plantation a few miles to the east; henceforth he never resided in Georgia again.[35]

It was sometime during the latter part of 1862 when Lee with his family moved to Knoxville, where he bought in January of the following year a property known as the Tennessee Hotel, consisting of twenty furnished rooms.[36] According to a statement he made five years after, he acquired also a nearby plantation of 150 acres, where he planned to set up an agricultural school. He had hardly got well established and raised a crop before he found himself in the midst of military operations, incident to the Federals' driving the Confederates out of Tennessee. In early September, 1863, Ambrose E. Burnside marched into East Tennessee and occupied Knoxville without a fight. The Federals seized Chattanooga also about this time, and soon thereafter the Battle of Chickamauga was fought. General James Longstreet, who had some part in that battle, had been detached from General Robert E. Lee's army in Virginia, and on his way back to Virginia had made an attempt to recover Knoxville in November. In the siege that followed, Daniel Lee found himself between the lines. His property had already been plundered by the Federal soldiers, who seized his livestock and ate up his provisions. In the fight his house and every other building on the place were burned to the ground, and as Lee complained, his family was "turned into the streets without food or shelter."[37] Here, it may well be assumed, Lee lost the mass of his personal archives, his library of books, and an accumulation of agricultural journals.[38] As he observed later, the war left him "little more than my thinking apparatus undestroyed."[39] It was a "political war," he said, "a most deplorable conflict," not yet ended in 1867 when he made these comments, "and may not end for generations to come."[40] This prediction might well have come true in a special sense, for a hundred years later the war was being "commemorated" through the years 1961–1965, when mock battles were being fought and a flood of new books about it were being written and old ones reprinted—all supposedly in a friendly feeling, at least on the surface. But during the preceding hundred years, if a Southerner did not mention the "late unpleasantness" to a visiting "Damn Yankee" then the Northerner would chide the "Rebel" for not having done so. If Lee had been living in 1965, he might well have predicted that the same situation would continue for the next hundred years.

Then, after Longstreet had marched back into Virginia and Lee had got his family settled in Knoxville again, he moved about unmolested for the remainder of the war. He visited his two sons on his farm in the District of Columbia near the end of the war. At this time he philosophized on the wastefulness of the American agricultural economy, hinting that such was one of the causes of the war. He took as his text "What a Rye Straw Teaches," holding in his hand a rye straw only eight and a half inches long from the tip of its roots to the top of the stem, which he had plucked from abused land where one seed could scarcely be produced from a square rod of land. "It teaches," he said, "the certainty of that supreme folly which refuses to study the natural laws that govern the productiveness of all farms. It teaches that hard work without knowledge leads directly to barrenness—no grain for head—to national discord and civil war. It places a father in the District of Columbia in this most unhappy position. One of his two sons may be drafted by the laws of Congress to fight in the federal ranks; the other may be conscripted by the popular sovereignty of eleven States and made to fight in the Confederate army. If either deserts or refuses to fire in battle, he will be shot. One brother kills the other. Tell me, kind reader, is the sovereignty of the Northern or Southern ballot-box able to do justice to the father and mother whose son had been slain in this republican manner?"

Lee continued, "[in] the county where the writer's family resides temporarily, farmers have been compelled to change masters five times in five months, swearing now to support the Confederacy and Mr. Davis, and now the Union, Mr. Lincoln and his abolition proclamations. These are only a few of the evils that come to America by natural sequence for the great common wrong of abusing their mother earth. A nation that has one hundred million acres of abandoned fields, and one hundred million more being rapidly impoverished provokes the judgment of an angry God. . . . The fruitfulness of the earth was created not for one, two or three generations only, but for all generations in all coming time. Hence the crime against humanity and the will of God to impoverish the soil of a continent."[41] As long as he lived, Lee would continue to preach the conservation of American farmlands and their proper uses; even war could not erase this preachment from his mind.

# ❧ XIII ❧

# Life in Tennessee—
# Knoxville and Nashville

AS LEE had been bitterly assailed in the North in antebellum times on account of his defense of slavery and as he had ridden out the war by staying in the South and not joining the Federal forces, it should not be surprising that he was to be remembered by Northern critics during the Reconstruction period. Soon after the war his name was brought into one of the most horrible of the "rebel atrocity" stories ever to be ground out by the Yankee atrocity mill. The story first appeared in the *Cincinnati Gazette*, and was copied by the *New York Times*, the *Washington Chronicle*, and other newspapers.

According to the story, Lee kidnapped or secured by the promise of higher pay a free Negro boy in Maryland and took him to Georgia as a slave and put him to work on his plantation near Athens. After the war had been going on for two years, Lee took him to Knoxville, Tennessee, and on to Jonesboro, in that state, where he hired him out and later had him put into a Confederate quartermaster unit as a teamster. Soon, the Negro boy, who was called Dick, ran away to get into the Federal lines, but when about twenty miles away, he was overtaken and brought back. To prevent his running away again, the hospital surgeon cut off both his legs above the knees, and for good measure, cut off both hands, leaving him to die. But the next morning, when the surgeon found him still alive, he remarked, "What, you d——d nigger, are you alive yet? I intended to kill you." He then dressed the wounds, and the boy was sent out into the country to a Negro hut, where he was found by the advancing Federal troops. Dick was sent to the Freedmen's Bureau, where he told this story, "which is believed by the officers of the Freedmen's Bureau," as the newspapers reported it.

Even with the repetition of this tale in the newspapers, it is difficult to understand why anyone even in the emotional state of mind following the close of hostilities could have believed it; but there may be some truth in the Hitler slogan that a lie big enough and repeated often enough would come to be believed.

One of the newspapers in Lee's old home town of Rochester, the *Union & Advertiser*, left some doubt that it believed the story, as was indicated by its headline: "Our Old Townsman, Dr. Lee, Implicated in Alleged Cruelty to a Negro—his Vindication." It then published a letter from William N. White—a well-known horticulturist of Athens, Georgia, but a native of New York, who had known Lee well for many years—giving the facts in the case. About the same time there was published in the *New York Times* another letter, from Lee himself. The truth as brought out in these letters apparently settled the matter in the minds of Northerners who were not so blind as those who would not see. The facts were that Lee, and later one of his sons, owned a farm in the District of Columbia; a few years before the war it had been rented to a free Negro, who let the rent accumulate for a few hundred dollars. When Lee pressed him for payment, the Negro suggested that his son Dick might go back with Lee to Georgia and "work out" the amount due. The Negro was glad to make this offer and get rid of his son, for he knew that Dick was a worthless fellow, who constantly got drunk and was frequently arrested and thrown into jail.

Lee, not knowing this, took Dick back to his Georgia plantation and set him to doing odd jobs. Dick soon showed what he was made of, for he shirked his work, got drunk, and got himself into jail, as was his custom back home, and Lee was charged with the expense of getting him out. Finally Lee was able to hire him to a neighboring planter, who soon found out that Dick was even worse than Lee had stated. Thereupon he turned him back on Lee, who now tried to finally rid himself of Dick by sending him back to the District of Columbia. Dick would not go, feeling that he was better off with Lee in Georgia.

After the war had been going on for about two years, Lee found himself still encumbered with Dick when he decided to move to Knoxville. In the fall of 1863, Dick voluntarily attached himself to a Confederate quartermaster unit as a teamster. Being unable to resist whiskey but able by hook or crook to get it, principally by stealing things with which to get it, Dick was generally found drunk as often as sober. On one wintry night he was found drunk, lying out in the open with both feet frozen. Lee reported in his letter, Dick was found dead drunk, "his feet were frozen, mortified soon, and both legs were amputated to save his life." After repeating in his letter the lurid horror story to refute it, Lee said in closing: "Will not the Cincinnati *Gazette* and other papers who have copied the above case 'of atrocious brutality, illustrative of slaveholding barbarism,' do me the justice to let their readers see the truth of the matter?"

Referring to the part of the horror story where the surgeon left Dick for the rest of the night after amputating both legs and arms, White said in his letter that it would have been impossible, because of the loss of blood, for Dick to have lived fifteen minutes unless his wounds had been dressed and properly bound up. White said further, "Dr. Lee, in fact, never had much to do with slave holding; his whole interest in the institution consisted of a single family, a man, his wife and children, employed more as house servants, and to these he was indulgent, maintaining not sufficient subordination to make them either useful or happy. The fact is they owned him."[1]

It seems that Lee was pestered by no more atrocity stories, and he was not greatly upset by this one. In fact, he was busy with other interests after he went to Tennessee; it was only the temporary Brownlow dictatorship of Tennessee that did upset him. One of those interests which he took on when he moved to Tennessee, incidental to his agricultural program, was dealing in real estate. In February, 1864, he bought a small lot of only a fourth of an acre in Knoxville, paying for it $2,000—in Confederate money doubtless.[2] Thereafter, almost to the year of his death, he bought and sold city lots and farm land. In 1866, he bought two lots in the city, near the Virginia and East Tennessee Railroad tracks,[3] and in 1873, he bought a house on a lot 58 by 100 feet for $600, which he sold later for the same price.[4] The next year he bought a house and lot for $250.[5]

Lee formed with H. E. Howarth a supply company; in 1874 his company took a mortgage from William Smith on his household and kitchen furniture in Franklin House, which was being used as a hotel. This document was to guarantee a debt of $331.55, which Smith had incurred with Lee's firm for provisions.[6] No doubt, some of these provisions came from farmlands which Lee had by this time acquired. By 1876 Lee seems to have disposed of most, if not all, of his Knoxville property, when he sold a small house and lot for $300.[7]

Dealing in city property seemed a little odd for a man of the land as Lee was; so, indeed, it was more in keeping with him to carry on transactions in farmlands. In addition to previous purchases of farmlands already noted, Lee, in 1867, bought twenty-nine acres

in the Twelfth Civil District of Knox County, which two years later
he sold back to James W. Thornton, from whom he had bought it,
for the same amount he had paid for it—$1,000, a transaction which
indicated that Lee had never paid for it at all.[8] In 1871, he bought
from Jefferson Griffin a tract of ninety acres on Mill Shoals Creek
for $484; and two years later, he bought sixty-five acres "on the
south of Holston River."[9]

Lee's principal land transactions were in the Fifteenth Civil District to the east of Knoxville. In 1865, he bought from James L.
Johnson for $1,500 a tract of 200 acres, for which apparently he was
unable to make payment, for the next year he returned the land
to Johnson for the same amount.[10] Later he bought and sold land
in this area, in varying amounts such as 42.5 acres, 50 acres, and
other acreages.[11]

The land which Lee bought early and held longest in East Tennessee was a tract of 478 acres that he bought on October 18, 1865,
from Allen R. Johnson for $1,500. It was, in reality, an exchange of
Lee's farm near Athens, Georgia, of 596 acres, for which on the
following October 25, Johnson paid Lee a like amount.[12] This farm
lay between Bay's Mountain and the French Broad River, north-
west of this long range of mountains and near the North Carolina
line. It was at the north end of Happy Hollow, a name which must
have indicated that the Lees had good neighbors.[13] The post office
was Gap Creek. The Lees moved there from their Knoxville home
in 1866.[14]

This was a rich, grassy, limestone country, all much to Lee's liking. Apparently there was a house on the property, but no barn—
at least none to meet Lee's needs or liking. So, he soon set to work
building one, forty by eighty feet, with sufficient stalls for cattle
and other livestock, as he was careful to pen his stock at night for
their protection as well as to garner their manure. Also, he built a
rat-proof corn crib.[15] On the place, were three apple orchards, a
large peach orchard, many grapevines, and a distillery "for making
apple, peach and grape brandy."[16] Besides eating peaches in season,
making pies, and preserving them, Lee had some of them dried,
giving "half for the drying." Their fermented juice made "a cheap
and pleasant drink," but "a vitiated public taste generally prefers
peach brandy in the South," he observed.[17]

Living in a limestone country, Lee produced some lime on his
place, and he recommended its production to others, remarking,
"I can burn one thousand bushels of the best stone lime at five cents
a bushel."[18] He was originally attracted to East Tennessee by noting

its limestone formation. He said that during the war he saw a man making saltpeter "at the mouth of a limestone cave a few yards from the Tennessee River" with limestone rock extending over the farm, and when he found that the owner "wanted Confederate money more than the land," he bought it.[19] This must have been the 150 acres that Lee said went with his first purchase in Knoxville, on which he intended to set up a farm school.

What attracted Lee more than saltpeter and limestone was the grassy sward he saw everywhere. Nothing was nearer his heart than grass and what could be made of it. Now he said less about the various grasses he had advocated when he lived in Georgia. Here in Tennessee he would depend on the native grasses and would reduce much of his farm to clover. He brought from his Athens, Georgia, farm clover seed which he sowed here with good results.[20] He was now away from the cotton economy and was glad of it.[21] Grasses and clover were basic for a livestock economy, and that was to be one of his chief interests from now on. "There is more money," he said, "in a crop of clover and a crop of fine mules than in a crop of cotton."[22] Another reason that got him away from other crops was the high cost of labor. In his latter days, he remarked that the Louisville and Nashville Railroad was paying $1.50 a day for common labor, which no farmer could pay. "Put more land in good pastures and stock," he advised, "and less under the plow." The plow was "the leading tool and servant of our old grass killing system of slavery farming, called planting."[23]

Nearing the last year of his life, he said that raising horses and mules for the cotton farmers had been "the best business" he had done "during the last twenty-five years."[24] Using the kind of expression with which he liked to strike home an idea, he said that raising grass and clover "and selling them in the shape of mules and horses had been our best business."[25] He did not use a stud establishment, but bought young horse and mule colts from neighbors and elsewhere for twenty-five dollars to thirty dollars a head and then sold them at a good profit when they were old enough to work.[26] As an example, he noted in 1868 that he had sold forty mules to purchasers in Arkansas for a profit of three times what they had cost him. He concluded that for a farm where stock was a chief consideration, there should be more than a sufficiency of grasses and clover for grazing so that there should be some left for mowing. Hay not needed for feed could be sold in Knoxville for twenty-five dollars a ton.[27] He recommended three acres of grasses and clover to one for the plow.

Lee had cattle which he used for home consumption but not as a money crop; but as for sheep he had long advocated them for not only Tennessee (where they were already well established), but also for the South as a whole. In 1872, he said that Georgia wool had recently been sold in Boston for sixty-two cents a pound. Raising a pound of wool could be done as cheaply as raising a pound of cotton, which sold for about a sixth as much. "Cotton culture should rest awhile," he said, "till sheep husbandry fertilizes the soil up to an average of one bale of cotton to the acre."[28] The South had room and potentialities for 100 million sheep. All it needed was to turn its lands into grasses. Referring to the tax which the radical Reconstructionists had placed on cotton, Lee said, "We live or rather stay under a government that taxes cotton culture, and pays a high bounty on wool growing by an almost prohibitory tariff on imported wool." Why not take advantage of the situation? "Let southerners raise wool as they can with the greatest facility," he said, "and northerners cotton by free negro labor to their hearts content." He recalled that his family back in New York had raised sheep, and he "had carried new born lambs in a sheepskin apron." Mutton was an excellent food, a value added to the wool clippings. He was not downgrading hog meat in favor of mutton; both were good. But, "if dogs kill sheep more than they do hogs, it is because dogs know more than some men and prefer the taste of mutton to that of swine's flesh."[29]

When Lee bought his East Tennessee farm, he found on it a gristmill run by an overshot waterwheel. Having long been advocating irrigation on farms, he used the power of this mill to "direct water to irrigate a pasture and meadow" and found the power thus applied much more profitable than grinding grain. This was something new and yet very old, for such had been done in Babylon, Thebes, Nineveh, and Sodom and Gomorrah—though it is difficult to see how such power could have been obtained at the latter two places, lying as they did in the flat plains of the Dead Sea. Lee gave directions how to irrigate sloping lands by the use of reservoirs filled by rainwater and by the use of ditches and gutters, some running downhill and others running with the contours. "Water that comes from the clouds," he said, "is an agricultural power of inestimable value; and nearly all of it may be utilized on a farm."[30] A hundred years later, the reservoirs which Lee advocated were being set up, but largely for fishing and swimming.

If there was no gristmill with its waterwheel on the place, then steam power could be used, as Lee said that it was used in England.

As previously noted, Lee argued that steam engines could pump water to reservoirs or through pipes directly to the fields; and coming down to specific details, he said that the energy in 112 pounds of coal could raise 715 tons of water 10 feet high, using "a six-horse power engine," enabling "a man and a boy to irrigate abundantly ten acres in as many hours."[31] Lee was almost as enthusiastic about water as he was about grasses. Surface water was filled with valuable chemicals that had been absorbed from the land, and all of it should not be allowed to "run unvexed to the sea." The human body was made up mostly of water, as was cotton, and the land must equally have its share. A century later Americans were also impressed with the uses and needs of water, which they were beginning to feel was not in sufficient supply. They had to contend with water pollution, which happily never was a problem or even thought of in Lee's day.

Lee's farm economy called for little hired labor, compared with a plow system; yet when he needed laborers he could get them for fifty cents a day. That was what he paid in 1869.[32] After the war, getting laborers was a problem for Southern planters, and to help solve it, Lee advocated labor-saving machinery, which he said was "more needed in our agriculture than anything else, to compensate in part for the loss of slaves."[33] This situation led him to comment on the freedmen as laborers. He said that a correspondent of the *Cincinnati Commercial* had visited a large Southern plantation in 1869 and reported that he had found that 140 freedmen had produced less cotton than 60 slaves had done before the war. In addition, the freedmen had injured or destroyed five times more working stock than the planter's slaves had done.[34] "There is no higher prince in America," Lee commented, "nor one more fickle, than an ex-slave of pure or impure African blood. The value of his race for farm servants is nearly destroyed by this elevation." Not only Southern agriculture had been destroyed, but "our civilization is in peril," he added. "Their temptations and conditions now, are vice, crime, contagious diseases and incurable laziness."[35] Lee cited a shining example to the contrary in the fact that Ben Montgomery had bought Jefferson Davis's old plantation in Mississippi and was making a success of it. Recently Davis had visited the plantation and had had a meal with Ben, "who waited on his old master affectionately." Lee added, "But whether the rising generation, reared generally with an excess of liberty, will do as well as their fathers and mothers, raised as slaves, I doubt."[36]

But what disturbed Lee as much or more than that the slaves who had been set free had become worthless laborers was that they had been made voters. He said that 4 million slaves had been freed and made into 800,000 "sovereign voters, who . . . can neither read nor write."[37] These comments Lee made soon after the war. In his very old age, in the late 1880s, he was equally pessimistic: "High taxes and Negro sovereignty have come without asking and apparently to stay."[38]

Lee was not a political animal, and apart from his short service in the New York legislature (and that solely to promote agriculture), he steered away from politics—except to comment on the pestilences of the times, brought on largely by politicians. "Politics is a muddy stream," he affirmed, "and those that bathe most in it will get soiled, if they are not drowned."[39] Congress wasted the national revenues on everything except what was important for the national economy—especially agriculture and agricultural statistics. It was "the machine of the industrial classes mainly." "Kind reader," he inquired, "can you tell me any practical way to place one small grain of common sense in the big head of a small man in Congress?"[40]

Not satisfied with the way the national government was managed, Lee was much less in agreement with the state government of Tennessee, particularly under the William G. Brownlow regime of 1865 to 1869. Why did Lee ever come to Tennessee or stay after Brownlow came into power? He answered that "if I could have obtained lime on my granite farm near Athens, Ga., at three times its cost to burn on this in Tennessee, I should not have made myself the unhappy subject of Brownlow's despotism, and that of men who have robbed me in my dwelling-house." These and others Brownlow pardoned out of the penitentiary to organize gangs of horse thieves and to help elect Ulysses S. Grant to the presidency in 1868.[41] These Brownlow gangs also burnt Lee's barn and called his brother out of bed one night and shot him dead, all "with impunity." In addition to burning down Lee's barn, they burned his distillery. Lee said that he would not rebuild his distillery, "nor run any opposition to Gov. Brownlow in the manufacture of 'hell-fire' in the loyal old county of Knox."[42]

In this Brownlow war against all former Confederates, the governor drove many people out of the state; but through thick and thin Lee chose to remain. This war was not so much "a sectional as a civil war, the latter being an infinitely more malignant and painful disease than the former." By high confiscatory taxes and the deprivation of the people's right to vote, hold office, serve on juries, and engage in other legal activities, Brownlow was driving people out of the state or to despair. Lee was forced to sell some of his property to prevent its being confiscated by Brownlow's high taxes. Lee sold farm land "on seven years credit" for a price that would have amounted only to rent. But the day of retribution was coming, Lee predicted, when "tens of thousands who now sow the winds with a vigorous arm will reap the whirlwind."[43] It was not long before Lee's prediction came true, when Brownlow and his henchmen ceased to rule Tennessee.[44]

Counting the Brownlow regime as only a temporary aberration, Lee could not forget that he was a loyal Tennesseean, who had come to stay. Therefore, it was not illogical in his mind to boost East Tennessee as a place next to the Garden of Eden. Long before he had thought of moving to Tennessee, he had remarked about his "intimate knowledge of the natural resources and advantages of East Tennessee."[45] He had much to say about the geology of this region, its limestone formations and otherwise.[46] It was a great grazing country, with its attractive valleys and mountains, where a person could raise a calf three years old more cheaply and with less herding and driving than a Texan could drive one to Knoxville or St. Louis. Mountain lands could be bought in large tracts at from $50 to $100 per thousand acres.[47] His neighbors in and about Happy Hollow and settlers in other parts of East Tennessee herded their cattle into the mountains to feed on wild peavines and other herbage, while the herdsmen amused themselves "by killing deer, 'bar' and other game, and catching mountain trout."[48]

The climate of East Tennessee was "exceedingly favorable to longevity." The heads of one of Lee's neighboring families had been married for sixty-two years and were still going well. Another neighbor "between eighty and ninety, cuts cord wood, splits rails, and works in the field as efficiently as a rugged farmer in the prime of life. Southern mountain air imparts the force of character which has given to this sequestered State three Presidents in less than forty years."[49] A visitor might look into the household of these hardy people and see "one female spinning flax on a little wheel, another spinning cotton, and a third spinning wool—all raised on the same farm." This was a diversity which Lee admired.[50]

In praising the opportunities in Tennessee, Lee was not inviting indiscriminately every "Tom, Dick, and Harry" Yankee to be coming to Tennessee or to the South. He was not opposing all Northerners coming, but he did not welcome Northern speculators in either land or politics. "There is an earnest effort making to persuade Southerners to part with their lands for about the price that Congress paid for their slaves."[51] It was not Northern capital or

Northern brains that the South needed most; but what it did need was to develop the great potentialities of its own brains. He said that the South's "toiling masses need capital in the brain—wealth in cultivated intellect—rather more than capital in gold or other property." Educate the people on whom the South must rely, and wealth would come as it had in the North. Lee said that he knew Tennessee families who could not read or write and who did not produce a hundred-dollar surplus over what they consumed, but with a proper education they would add a thousand dollars to their income.[52]

Before moving to Tennessee, Lee had been editor of the *Genesee Farmer*, the *Southern Cultivator*, and the *Southern Field and Fireside*; he had been a contributor to various other farm journals. With his facile pen and a mind bubbling over with information he wanted agricultural people to have, it would hardly be logical to think that he would suddenly fall silent. The war had its effect on him, but it could hardly close him up. What those East Tennessee mountain families needed most was to take farm journals and be able to read them. There "isolated human families and hybernating bears make equal progress and equal improvement from one generation to another."[53]

After moving to Tennessee, Lee did not forget Athens, Georgia, nor did Athens forget him. In January, 1870, some of his old friends and acquaintances there decided to start another journal called the *Farmer and Artisan*—the *Southern Cultivator* was already being published there, having moved from Augusta to Athens about the end of the war. The fact that Lee agreed to be the "Principal Editor" seems to have been the main reason for adding it to the expanding list of Southern farm journals. Lee thought that it had a right to a place because it was to be a weekly, and would thereby bring fresher information and more of it to the farmers. In taking hold, he said, "Every well informed farmer and mechanic knows there is a black cloud of ignorance and folly brooding over our fair domain which must be dispelled for the common benefit of all." In announcing its publication and Lee's connection with it, the proprietors said, "Though his home has been twice desolated by fire—once by the Federal army and once by accident—he is full of vigor and hopeful confidence in the future of the South, and his contributions to these pages evince the grace and vigor which have rendered him a favorite with the reading public."[54] The bright hopes of this journal were blasted, for it had a short life.

After a struggle for two years, the *Farmer and Artisan* was merged with the *Plantation*, a farm journal established in Atlanta in the very month of the birth of the *Farmer and Artisan;* it was also a weekly for the first two years. It ran as its subtitle: "Devoted to the Interests of Agriculture, Rural Economy, and the Benefit of the People." Its editor was Charles W. Howard, a well-known Georgia reformer, who in 1859 had founded and edited the *South Country-man* for a short time. In 1872 Howard resigned on account of ill health and the proprietor, Benjamin C. Yancey, immediately hit upon Lee as the perfect successor, referring to him as "an old soldier in the field of Rural Literature."[55] The *Plantation* continued for only one year longer than the *Farmer and Artisan* had run, the last known issue being in 1873. So Lee was left without any editorial duties until later in life, but his pen was not stilled, for he continued to write for farm journals, both South and North. Occasionally he sent a contribution to the *Southern Cultivator,* the Albany (New York) *Cultivator,* the Albany *Country Gentleman,* the *Albany Cultivator & Country Gentleman* (a combination of the two), the New York *American Agriculturist,* the Rochester *Rural New Yorker,* the Baltimore *American Farmer,* and others. Of course, it should be evident that he did not move to the place of origin of the journals which he edited, and naturally not to those to which he was a contributor. Going neither to Athens, where he edited the *Farmer and Artisan,* nor to Atlanta, where the *Plantation* was published, he remained on his East Tennessee farm until the latter part of 1872, when he decided to remove to Nashville, for reasons which he seemed never to have published—and there he remained until his death. At the time of his move, Lee referred to himself as "an old and independent thinker."[56]

In December, 1866, there appeared in Nashville a new farm journal with an old name—the *Tennessee Farmer.* In antebellum times a journal with that name had been edited in Knoxville but had long ceased to exist. With Daniel Lee now in Nashville, right under the nose of this new journal, it was inevitable that he should begin writing for it and soon become a "Contributing Editor." His first contribution appeared in February, 1887, and from then on for a few years he wrote voluminously for it and soon began promoting it as vigorously as he had the *Genesee Farmer* or the *Southern Cultivator.* He wanted *Tennessee Farmer* clubs organized in every county in the state, not only to take the journal but also to promote the programs he was advocating.[57] On reading Lee's first article in the journal, an old Knoxville acquaintance who had lost track of his old friend, said, "It was a pleasant surprise for me to learn that Dr. Daniel Lee is still alive and well. There are but few if any living to whom the farmers of the United States are more indebted."[58]

Not being content with being a contributing editor to one farm journal, Lee in 1888 joined the editorial board of his old *Southern Cultivator* which he had edited for a dozen years before the war. For more than a dozen years it had forgotten him, for it had not published a line from him during that time. But now William L. Jones, one of the editors of this journal, had resigned to become a professor of agriculture in the University of Georgia, leaving a vacancy that Lee filled. In a flattering announcement the journal referred to him as the "venerable but still active Dr. Lee." There was a "vitality and progressiveness about him, physically and mentally," that showed that he might wear out at least but never rust out or become an old fogy. He returned to his old editorial job "better fitted than at any previous period" of his life to guide farmers forward to prosperity.[59] He was internationally known and stood "in the front ranks of agricultural writers by reason of long experience and profound thought."[60] Even some of his old Northern acquaintances were impressed. A writer in the New York *Weekly Star* said, "Though an octogenarian the doctor is still a vigorous and versatile writer on the practice and science of agriculture."[61]

Both as editor and contributor over the past fifty years and more, Lee could hardly be writing something new during the last dozen and a half years of his life except on timely topics. But he could and did write as well on his old themes: artificial manures, farm labor, soil exhaustion and restoration, grasses, wheat, the wealth being lost in disposal of cotton seed, promotion of livestock, forestry, and so on. He tried to form an association for the promotion of grass culture and stock raising, and in an attempt to protect sheep raising by taxing sheep-killing dogs (all dogs being potentially so), he snarled at the Tennessee legislators for not passing the necessary laws: "They snap their canine teeth and every time for the dogs."[62] His old themes were worth playing over again and again to impress them on those who had heard them before and to inform new generations as they came along. Also, he could fill his columns by taking up something which somebody else had written, commenting on it, agreeing or disagreeing.

But getting away from his old beaten paths, he could find plenty to write about on the happenings of the times, pestilential and otherwise; and it should be noted that although he was in disagree-

ment now and then, he was fundamentally an optimist and a friendly writer most of the time. He wrote about Henry Ward Beecher's
trial and thought that it had not helped Christianity very much;
with the coming of harnessing electricity, he thought that there
were uses for it on farms; he commented on the power of the press
and the rise of labor organizations and strikes; and he could in an
angry mood comment on taxing the poor man's necessities such
as salt and sugar and letting the rich man escape to increase his
fortune. The rich benefited "by this system which descends in
taxing the throats of sixty million people every day of their lives
to make a few men rich enough to buy a controlling power" in
government.[63]

He had thoughts on railroads, especially about the narrow gauge,
which could operate a third cheaper than the standard gauge, and
thereby make a third lower rates in getting farmers' products to
market. Even if not built in long lines, these railroads could be
made feeders to the standard gauge roads already in operation.[64]
Refrigerator cars, now coming into use, greatly attracted Lee. An
"ice-house on wheels," he called them, and said that they were
"something new under the sun." Farmers could now get perishable
products to market.[65] The great Chicago and Boston fires reminded
him that almost every day some country residence burned down.
He told how to avoid these disasters by proper construction inside
and out. He said that it was "much cheaper to prevent burning
down a house than to build a new one."[66]

In the 1880s, the appropriation of Africa by the nations of Europe
was going apace, and this put thoughts into Lee's head. He said
that the dark continent was fast becoming civilized and "a white
man's territory." Cities were being built, railroads constructed,
telephones installed, and all else that went with the white man's
civilization. What did this mean to the Southern farmer? It meant
that since much lumber would be needed there would be a market
for the Southern product, and thereby the people would get some
compensation for the emancipation of their slaves "by selling timber and lumber to go to Africa and aid in the development of a new
civilization."[67]

When Lee moved to Nashville he did not immediately divest
himself of all of his Knox County property, although in the year of
his going (1872), and, perhaps, preparatory to his departure, he
sold 160 acres in the Fifteenth Civil District for ten dollars an
acre.[68] Later (in 1879), he sold 20 acres of his original home farm
for fifteen dollars an acre,[69] and five years later he sold 160 acres

(presumably not part of his home farm) for about nine dollars an acre. This land was to be paid for in fourteen annual installments, and to secure these payments (and as part payment) Lee was to receive two broodmares, two cows and three calves, "and other cattle, sheep and swine." Also he was to receive 200 bushels of corn annually and the hay and fodder grown on the place "to keep the stock of the same."[70] In 1887, Lee disposed of land in the Thirteenth Civil District bordering on "the river bank" for $800.[71] He owned throughout the remainder of his life part of his old home farm, for he noted in 1888, being now more than eighty years old, that he had walked more than ten miles over the place.[72]

True to his nature, Lee had hardly reached Nashville before he was out over the countryside looking at farmlands and visiting stock farms and country estates. Inevitably his footsteps were soon directed to the famous Belle Meade, the "Queen of Tennessee Plantations," then owned by General William Giles Harding (1808–1886). After meeting the general, Lee, accompanied by him, was soon walking over a part of this 3,500-acre farm and plantation.[73] Soon he was visiting the Hermitage, the Andrew Jackson estate; thinking of farm economy more than history, he suggested building it into an "industrial university."[74]

Equally true to Lee's nature, he must have a farm of his own, for he never stayed at a place long before he acquired one. In 1874, he came into possession of a seventy-four-acre farm, and soon he was buying other farmlands.[75] Also, he bought and sold city property.[76]

Despite Lee's advanced age—he was now an octogenarian—he would travel long distances to make speeches on the subjects of agricultural reform. When he was eighty (or eighty-six, depending on the uncertainty of his birth year), he was invited to speak at the Farmers' Inter-State Encampment in Spartanburg, South Carolina. In his speech to the "farmers' encampment," he was asked to outline the needed developments in Southern agriculture, a subject which no man was "better fitted to discuss" than Lee. In mentioning his passing through Atlanta on his way to Spartanburg, a local newspaper referred to him as "the venerable, but still active Dr. Daniel Lee."[77] Introduced by former governor Johnson Haygood, Lee made an address satisfactory to all of his hearers.[78]

The last years of Lee's life were made up farming a little; writing for farm journals—especially for the *Tennessee Farmer* and *Southern Cultivator;* buying and selling some real estate in the city and county; and visiting among his children, grandchildren, and great-grandchildren. He was now living with one of his children (his wife Sarah died a year or two before his own death).[79] Earlier, and probably one of the reasons for his moving to Nashville, was the offer and his acceptance of a clerkship in the Tennessee Department of Agriculture while Joseph Rucker Killebrew was the secretary. Apparently he did not last long in that position, for Killebrew remembered later when he wrote his "Reminiscences" that Lee was "totally inefficient on account of age and infirmity. He had lost his power and could not keep up with the rapid movements of the times." [80] There must have been some incompatibility between the two men, for others who commented on Lee at this time held him to be highly vigorous in both mind and body.

Lee died early in 1890 (probably sometime in March or before— certainly not later), and it might be said that literally "he died in harness," for in October, 1889, he had an article in the *Southern Cultivator*, which, although written in proper English construction, rambled over several unrelated items before it ended. In it he told of the advantages the South had in raising apples for use at home and for export to Europe; the way that flour could be best preserved in glass containers rather than in barrels; the great advantages of Southern climate; and the uses of grasses, one of his favorite topics.[81]

Noting that his age was eighty-eight, the *Southern Cultivator*, after paying its last respects to this "venerable, honored and distinguished editor," quoted a eulogy of him that appeared in the *Atlanta Journal*: "Venerable in years and feeble in body. . . . After a career of useful devotion to the agricultural interests of the South, this modest yet eminent scientist has gone to his final reward, leaving a host of friends and no enemies to revere his memory. . . . He lived a busy but unambitious life, and his work for the farmers of the South will remain for their guidance long after his body shall have mouldered to dust." [82]

Lee was so unconcerned with death that apparently he never thought of making a will; so, to settle his estate the court appointed the Nashville Trust Company. According to the brief records of its administration, he left effects worth slightly less than $4,000 and the findings ended with the customary statement: "The within is a full and perfect inventory of all the goods and chattels, rights and credits of the estate of Daniel Lee, deceased, which have come to our hands, knowledge or possession to the best of our belief." [83]

In addition to what the *Atlanta Journal* obituary had to say, it might still be asked: Really, what sort of man was Daniel Lee? It is evident that Lee was an unusual man, a largely self-educated and self-made man, who had a deep and lifelong passion for the land and

for those who lived on it and made their living from it; that he was
a sincere man free from all guile and sectionalism, but who upheld
the South before, during, and after the Civil War from principles
which he had always had and which he brought with him from the
North; that (assuming his age was eighty-eight at his death) having
lived forty-seven of those years in the North and the last forty-one
in the South, he was broadminded enough to think of himself as
an American—not a Northerner or a Southerner; that his agricultural program was sound, and his scientific knowledge was based on
experimentation and not guesswork; and that much of what he
stood for was to be realized in the twentieth century.

Lee was a religious man though it is not known that he belonged
to any organized denomination. He was a facile and versatile writer,
with a mind generating thoughts faster than his pen could take them
down. He was well read in history, even into ancient times, and he
did not neglect reading scientific works. He was especially interested in the chemistry of soils, their depletion and restoration, which
led him into a study of geology; he often argued for grass economy
against the plow; and he never ceased to advocate diversification in
agriculture. He made a competent living for his family and always
had a sufficient amount of money at his command to take care of his
needs. Being carried away by his enthusiasm for some good cause,
he was impractical in his attempts to raise money for its promotion.
This enthusiasm led him to exaggerate in some of his statements.

It is somewhat a mystery that a man of his parts should have been
so thoroughly forgotten and hardly ever to have been mentioned
after his death; even more mysterious is the fact that the principal
Nashville newspaper, the *Banner*, did not even record his death,
which took place in a city where he had lived the last seventeen
years of his life.[84]

It is not as much a mystery, however, why he should have been
ignored and forgotten in the North, because many there felt that
he had deserted his principles when he went South and that he had
joined the slavocracy and had toadied to that cult for what he
could get out of it. Even close associates during his Northern residence entirely ignored his death. The *Cultivator & Country Gentleman*, with which Luther Tucker and John J. Thomas were
connected, never mentioned Lee's name, and even when the *Genesee Farmer* went out of existence in 1865, in giving a summation of
its long history, it was able to skirt even mentioning Lee's name,
although he had for many years been associated with that journal
and for some years had edited it and owned it.

If Lee must be judged by what many people know and feel in the twentieth century relative to such outmoded institutions as slavery and other Southern doctrines of those times, and of attitudes toward the Civil War and Reconstruction, then Lee would be put down as a narrow racist and sectionalist; but if he be judged by the spirit of the times and places wherein he lived, then he should be rated as an admirable character, modest, unambitious, scientifically minded, and dedicated to helping the farmer and the common man.

# Notes

CHAPTER I

1. Brief incomplete sketches of Lee may be found in Albert L. Demaree, *The American Agricultural Press, 1819–1860* (New York, 1941), 67n. and Alfred G. True, *A History of Agricultural Education in the United States, 1785–1925*, United States Department of Agriculture Miscellaneous Publications, no. 36 (Washington, D.C., 1929), 72.

2. Demaree in his *American Agricultural Press*, 67n., gives Lee's birth year as 1802. It is not known where he got this information. The United States Census, 1860, Georgia, Cass-DeKalb, Population (MS microcopy T-7, roll 27), 130 microfilm in General Library, University of Georgia, gives Lee's age as fifty-two, which would make his birth year 1808. Demaree probably found the better information than what was given in the census, owing to mistakes in transcribing in the census office. See also *The Plantation. Devoted to the Interests of Agriculture, Rural Economy and the Benefit of the People*, III (August 14, 1872), 519.

3. *The Southern Cultivator, A Monthly Journal, Devoted to the Interests of Southern Agriculture...*, XXVI (May, 1868), 131.   4. *Ibid.*, XLVI (November, 1888), 515.

5. Augusta *Weekly Constitutionalist*, February 23 (1,6), 1859; March 23 (3,1), 1859. For a discussion of slavery in New York, see Edgar J. McManus, *A History of Negro Slavery in New York* (Syracuse, N.Y., 1966).   6. *Plantation*, III (September 25, 1872), 609.   7. *Ibid.* (May 22, 1872), 329.   8. Athens *Southern Watchman*, May 10 (3, 1–2), 1860.

9. *Plantation*, III (July 17, 1872), 455; *Southern Cultivator*, XXV (September, 1867), 281; XLVII (April, 1889), 181; *Tennessee Farmer*, I (February 24, 1887), 1; *The Cultivator & The Country Gentleman: The Farm, the Garden, the Fireside...* (January 26, 1871), 51. In 1867, Lee wrote, "After buying cows and calves at $5 a head in Southern Illinois (not counting the calves,) building mills on the Little Wabash, and suffering much sickness, I sought and found a healthier country farther South." It is hard to reconcile this statement with any period of Lee's life, unless in his early twenties before he went to Chautauqua County to practice medicine, and it would have been a quarter century thereafter before he went South. See *ibid.* (November 14, 1867), 316. For quotation in main narrative, see *Rochester Daily American*, January 29 (2, 4), 1853.

10. *Southern Cultivator*, XXIV (December, 1866), 280.   11. *Ibid.*, XLV (June, 1887), 251; *Southern Field and Fireside*, I (November 5, 1859), 190.   12. *Plantation*, III (May 22, 1872), 329.   13. *The Cultivator: A Monthly Journal, Devoted to Agriculture, Horticulture, Floriculture, and to Domestic and Rural Economy...*, I (February, 1844), 47.   14. *Southern Cultivator*, XVI (July, 1858), 201–2.   15. *Ibid.*, XLVII (July, 1889), 321.

16. *Tennessee Farmer*, I (February 24, 1887), 1; *Southern Cultivator*, XLVI (June, 1888), 257; XLVII (April, 1889), 181. At one time he had been employed by the New York State Agricultural Society to deliver lectures on agricultural chemistry in "every county" in the state. *Report of the Commissioner of Patents for the Year 1845* (House Document, no. 140, 29th Cong., 1st Sess.), 504; *The Genesee Farmer: A Monthly Journal Devoted to Agriculture & Horticulture, Domestic and Rural Economy. Illustrated*

with *Engravings of Farm Buildings, Domestic Animals, Improved Implements, Fruits,* etc., XII (1851), 82.

17. Albany (N.Y.) *Cultivator*, I (February, 1844), 47; (March, 1844), 76; (June, 1844), 185; *Plantation*, III (August 14, 1872), 513; *Southern Cultivator*, XLVI (February, 1888), 71. 18. *The Country Gentleman: A Journal for the Farm, the Garden and Fireside* . . . , XI (April 15, 1858), 234; *Plantation*, III (September 11, 1872), 577. 19. True, *History of Agricultural Education*, 50; *Southern Cultivator*, XLVI (May, 1888), 211; Albany (N.Y.) *Cultivator*, I (February, 1844), 47; (April, 1844), 110–11; *Genesee Farmer*, VI (1845), 83.

20. *Genesee Farmer*, VI (1845), 99. 21. *Ibid.*, 18. 22. *Ibid.*, 178. 23. All quotations from Lee's report are found *ibid.*, 69–71. 24. *Ibid.*, 18. 25. *Ibid.*, 100. 26. *Ibid.*, 96. 27. *Ibid.*, 99. 28. *Ibid.*, VIII (1847), 16. 29. *Ibid.*, XV (1854), 321. 30. *Ibid.*, VIII (1847), 9. 31. *Ibid.*, X (1849), 178. 32. *Ibid.*, VI (1845), 69. 33. *Ibid.*, VII (1846), 28, 103, 202, 243; VIII (1847), 10–11, 21, 24, 32; *Southern Cultivator*, XLVII (February, 1889), 85; Demaree, *American Agricultural Press*, 67n.

34. *Genesee Farmer*, VII (1846), 274. 35. *Ibid.*, 199. 36. *Endowment of the Terrell Professorship of Agriculture, in the University of Georgia* (Athens, Ga., 1854), 15; Athens *Southern Watchman*, May 10 (3, 1–2), 1860; *Plantation*, III (May 22, 1872), 329.

37. Demaree, *American Agricultural Press*, 339. 38. *Southern Field and Fireside*, III (September 7, 1861), 127.

### CHAPTER II

1. *Southern Cultivator*, V (January, 1847), prospectus bound with this volume. 2. *Ibid.*, V (August, 1847), 120. 3. *Ibid.* Lee's name was not listed as editor until the December number. 4. *Ibid.*, VII (May, 1849), 72. 5. Quoted in Demaree, *American Agricultural Press*, 374. 6. *Southern Cultivator*, XLVI (June, 1888), 257. 7. *Ibid.*, 254; *Genesee Farmer*, X (1849), 82. 8. *Genesee Farmer*, IX (1848), 120. 9. *Ibid.*, 37–38.

10. *Southern Cultivator*, XLVI (1888), 254. 11. Augusta *Weekly Constitutionalist*, March 2 (3, 1), 1859; *Southern Cultivator*, VI (February, 1848), 24. 12. *Southern Cultivator*, VI (May, 1848), 72; (August, 1848), 123; XLVI (July, 1888), 306. 13. *Genesee Farmer*, IX (1848), 9. 14. *Southern Cultivator*, XV (November, 1857), 333. 15. *Genesee Farmer*, X (1849), 9.

16. *Southern Cultivator*, XLVII (October, 1889), 501. 17. *Genesee Farmer*, XII (1851), 123. 18. *Ibid.*, XIII (1852), 224. 19. *Ibid.*, VIII (1847), 84. 20. *Ibid.*, XI (1850), 209.

21. *Ibid.*, IX (1848), 118. 22. *Ibid.*, XII 1851), 9. 23. For instance, *ibid.*, VIII (1847), 16. 24. *Ibid.*, 9. 25. *Ibid.*, X (1949), 10.

26. *Southern Cultivator*, XLVII (January, 1889), 5–6. 27. James C. Bonner, *A History of Georgia Agriculture, 1732–1860* (Athens, Ga., 1964), 122–23. 28. *Report of the Commissioner of Patents for the Year 1849* (Washington, D.C., 1850), Part II (Agriculture), 262–63. 29. *Southern Cultivator*, XII (March, 1854), 74. 30. For instance, Augusta *Weekly Constitutionalist*, March 2 (3, 1), 1859.

31. *Southern Cultivator*, VII (January, 1849), 9; (October, 1849), 145. 32. *Ibid.*, XII (January, 1854), 10. 33. *Ibid.*, XV (May, 1857), 147. 34. *Ibid.*, XIII (January, 1855), 9.

35. *Southern Field and Fireside*, I (August 27, 1859), 110.

36. *Southern Cultivator*, XIII (February, 1855), 43. See also *Country Gentleman*, XI (April 15, 1858), 234–35. 37. Augusta *Weekly Chronicle & Sentinel*, January 3 (2, 6), 1848. 38. *Southern Cultivator*, VIII (March, 1850), 40. 39. Augusta *Weekly Chronicle & Sentinel*, May 30 (1, 2), 1849.

40. *Southern Cultivator*, XV (July, 1857), 203. 41. *Southern Field and Fireside*, I (July 30, 1859), 76. 42. *Southern Cultivator*, XXVII (May, 1869), 172. 43. *Plantation*, III (August 7, 1872), 502. 44. *Southern Cultivator*, XI (September, 1853), 284.

45. *Plantation*, III (August 7, 1872), 502.

CHAPTER III

1. Demaree, *American Agricultural Press*, 375.   2. *Southern Cultivator*, VIII (May, 1850), 73 (for quotation); IX (1851), 184.   3. Demaree, *American Agricultural Press*, 336–39.

4. *Genesee Farmer*, VIII (1847), 274.   5. *Ibid.*, X (1849), 82.   6. *Ibid.*   7. *Ibid.*, XI (1850), 202.   8. *Ibid.*, IX (1848), 37.   9. *Ibid.*, X (1849), 178.   10. *Ibid.*, IX (1848), 214.   11. *Ibid.*, VIII (1847), 274.   12. *Ibid.*, IX (1848), 216.   13. *Ibid.*, 165–66.   14. *Ibid.*, 213.   15. *Ibid.*, 14.   16. *Ibid.*, X (1859), 249.   17. *Ibid.*, XV (1854), 304.   18. *Ibid.*, XI (1850), 52.   19. *Ibid.*, XII (1851), 10, 30.   20. *Ibid.*, X (1849), 273.   21. *Ibid.*, 274.   22. *Ibid.*, 273, 274.

23. *Ibid.*, 274. There is some confusion here. The printing plant that Lee said he had purchased evidently was the establishment the Jerome Brothers had owned and used to print their *Rochester Daily American*, several other newspapers, and the *Genesee Farmer*. In 1850 they sold everything, including the *American*, to Lee and two other purchasers.

24. *Ibid.*, VIII (1847), 273; IX (1848), 285; XII (1851), 124; XIII (1852), 97.   25. *Ibid.*, XI (1850), 100.   26. *Ibid.*, XIV (1853), 332.   27. *Ibid.*, 330, 331.

28. The last issue of the *Genesee Farmer* was for December, 1865. It was sold and combined with the *American Agriculturist*, published in New York City.

29. *Genesee Farmer*, XII (1851), 275.   30. *Ibid.*, XIII (1852), 329.   31. *Country Gentleman*, XI (February 18, 1858), 107.   32. *American Agriculturist*, IX (February, 1850), 66.   33. *Genesee Farmer*, X (1849), 251–54, 277–80.

34. *Southern Field and Fireside*, I (December 31, 1859), 254. See also, Daniel Lee, Washington, D.C., December 4, 1851, to Thomas Ewing, in Thomas Ewing Family Papers, Manuscripts Division, Library of Congress, Box 55. This note was furnished me by Col. H. B. Fant of the National Archives.

35. Augusta *Weekly Chronicle & Sentinel*, October 27 (3, 2), 1847.   36. *Ibid.*, October 27 (2, 8), 1847.   37. *Ibid.*, October 4 (3, 3), 1848.   38. *Ibid.*, August 29 (3, 2), 1849.

39. *Ibid.*, August 22 (1, 2), 1849.   40. *Ibid.*, August 22 (1, 6), 1849.   41. *Ibid.*, October 10 (1, 2), 1849.   42. *Ibid.*, April 4 (3, 2), 1849.   43. *Ibid.*, April 18 (3, 4), 1849. See also *ibid.*, August 22 (1, 6), 1849.   44. *Ibid.*, August 10 (1, 2), 1849.   45. *Southern Cultivator*, VII (December, 1849), 185.

46. *Ibid.*, XVII (June, 1859), 176. The seven positions Lee held during the years 1847–1862 were editor, *Genesee Farmer*, 1845–1854; agricultural editor, Augusta *Chronicle & Sentinel*, 1847–1849; editor, *Southern Cultivator*, 1847–1859; editor, Patent Office reports, 1849–1853; editor, *Rochester Daily American*, 1849–1854; professor, University of Georgia, 1855–1862; and agricultural editor, *Southern Field and Fireside*, 1859–1862.

47. Augusta *Weekly Constitutionalist*, April 13 (4, 2), 1859.


CHAPTER IV

1. Augusta *Weekly Chronicle & Sentinel*, November 21 (3, 2), 1849.   2. Paul W. Gates, *The Farmer's Age: Agriculture, 1815–1860*, vol. III, *The Economic History of the United States* (New York, 1966), 330–36; *The Yearbook of Agriculture, 1940. Farmers in a Changing World* (Washington, D.C., 1940), 246–47.

3. Quoted in Gates, *Farmer's Age*, 332.   4. *Southern Cultivator*, VI (July, 1848), 104.   5. Augusta *Weekly Chronicle & Sentinel*, August 1 (3, 3), 1849.   6. *Ibid.*   7. *Ibid.*   8. *Ibid.*

9. *Patent Office Report, 1849*, II, 5; Daniel Lee, Washington, D.C., December 4, 1851, to Thomas Ewing, in Thomas Ewing Family Papers, Manuscripts Division, Library of Congress, Box 55. This note was furnished me by Col. H. B. Fant of the National Archives.   10. *Southern Cultivator*, VII (December, 1849), 185.   11. Augusta *Weekly Chronicle & Sentinel*, December 5 (2, 3), 1849.   12. *Ibid.*, November 21 (3, 2), 1849.

13. *Southern Cultivator,* x (January, 1852), 1. 14. *Patent Office Report, 1849,* II, 6–49, 207–15, 231–42, 261–65, 304–5, 307–13; *ibid., 1850,* II, 25–82, 118–20, 145–49; *ibid., 1852,* II, 1–23; *Southern Cultivator,* XI (March, 1853), 72. 15. *Genesee Farmer,* XIII (1852), 35.

16. *Patent Office Report, 1849,* II, 6, 83–86; *ibid., 1851,* II, 129–31; *ibid., 1852,* II, 58–60. 17. *Ibid., 1849,* II, 12; Demaree, *American Agricultural Press,* 57. 18. *Southern Field and Fireside,* III (July 27, 1861), 79. 19. *Patent Office Report, 1849,* II, 10, 12, 14, 15; *Southern Field and Fireside,* II (June 7, 1860), 262. 20. *Genesee Farmer,* XIV (1853), 272.

21. *Southern Cultivator,* XLVII (October, 1889), 449. 22. *Ibid.,* XLVI (June, 1888), 25. 23. *Genesee Farmer,* XIV (1853), 265–66. 24. *Patent Office Report, 1849,* II, 9.

25. *Genesee Farmer,* XIV (1853), 273. 26. *Patent Office Report, 1849,* II, 9. 27. *Genesee Farmer,* VII (1846), 247; XIV (1853), 207–8, 311–13. 28. Augusta *Weekly Chronicle & Sentinel,* April 18 (2, 6), 1849; *Southern Cultivator,* XI (September, 1853), 283.

29. *Patent Office Report, 1845,* 217, 489; *ibid., 1849,* II, 261–65.

30. *Country Gentleman,* XXV (January 5, 1865), 12; (April 14, 1865), 233; *Southern Cultivator,* XIII (January, 1855), 14; (September, 1855), 283; XVII (May, 1869), 172; *Plantation,* III (May 8, 1872), 296; (August 14, 1872), 513; *Genesee Farmer,* XIV (1853), 332; *Rochester Daily American,* January 29 (2, 4), 1853. 31. *Journal of the United States Agricultural Society,* I (August, 1852), 26–27. 32. *Ibid.,* 24, 28–29. 33. *Southern Cultivator,* x (March, 1852), 65. See also *ibid.* (April, 1852), 97–100. 34. *Genesee Farmer,* XII (1851), 36. 35. *Journal of the United States Agricultural Society for 1854,* 12.

36. *Genesee Farmer,* XII (1851), 177–78; XIV (1853), 302; XV (1854), 36; *Plantation,* III (July 3, 1872), 425.

## CHAPTER V

1. *Genesee Farmer,* XII (1851), 36. 2. *Ibid.,* XIII (1852), 84. 3. *American Farmer,* VII (August, 1851), 70. 4. *Journal of the United States Agricultural Society,* I, 5–6; *Southern Field and Fireside,* I (May 19, 1860), 414. 5. *Genesee Farmer,* XIV (1853), 191; *Southern Cultivator,* IX (August, 1851), 121–22; IX (October, 1851), 152–53.

6. *Genesee Farmer,* XIII (1852), 235–38; *Journal of the United States Agricultural Society,* I, 6. 7. *Journal of the United States Agricultural Society,* I, iii–iv, 7; II, 13–14, 24; *Genesee Farmer,* XIII (1852), 237; *Southern Cultivator,* x (August, 1852), 225–26. 8. *Journal of the United States Agricultural Society,* I (second number n.d.), 19; *ibid.,* for 1854, p. 255; *Patent Office Report, 1852,* II, 22; *Genesee Farmer,* XV (1854), 34–35; *Southern Cultivator,* XI (April, 1853), 108. 9. *Southern Cultivator,* XI (July, 1853), 207; (September, 1853), 273; *Genesee Farmer,* XV (1854), 26. 10. *Tennessee Farmer,* February 24, 1887, 4.

11. *Southern Cultivator,* XI (March, 1853), 71; *Genesee Farmer,* XIV (1853), 208, 300. 12. Augusta *Weekly Constitutionalist,* March 23 (1, 2), 1859. 13. *Southern Field and Fireside,* I (December 31, 1859), 254. 14. *Southern Cultivator,* XVI (February, 1858), 58–60. 15. *Genesee Farmer,* XIV (1853), 208. 16. *Ibid.,* 269–70. 17. *New-York Daily Tribune,* December 10 (4, 5), 1852. 18. *Ibid.* 19. *Ibid.,* December 11 (4, 5); *Rochester Daily American,* January 25 (2, 4), 1853. 20. *Southern Cultivator,* IX (February, 1851), 24; Augusta *Weekly Chronicle & Sentinel,* April 9 (2, 7), 1851.

21. *Rochester Daily American,* January 27 (2, 4), 1853.

22. *Southern Cultivator,* IX (February, 1851), 24. 23. *Ibid.* 24. *Ibid.* 25. *Ibid.,* 25; Augusta *Weekly Chronicle & Sentinel,* April 9 (2, 7), 1851. 26. *Rochester Daily American,* January 10 (2, 2), 1853; Augusta *Weekly Chronicle & Sentinel,* April 9 (2, 7), 1851; *Southern Cultivator,* IX (February, 1851), 24. 27. *Southern Cultivator,* IX (August, 1851), 121. 28. Augusta *Weekly Chronicle & Sentinel,* April 9 (2, 7), 1851. 29. *Southern Cultivator,* IX (February, 1851), 25.

30. *Rochester Daily American,* January 29 (2, 4), 1853. 31. Quoted, *ibid.,* Decem-

ber 10 (2, 3), 1852.  32. *Ibid.*, January 25 (2, 4), 1853.  33. *New-York Daily Tribune*, December 10 (4, 4), 1852.  34. *Ibid.*, December 11 (4, 5), 1852.  35. *Ibid.*, December 10 (4, 4), 1852.  36. *Ibid.*, December 10 (4, 3), 1852.  37. *Rochester Daily American*, December 29 (2, 3), 1852.  38. *New-York Daily Tribune*, January 1 (4, 6), 1853.

39. *Rochester Daily American*, January 10 (2, 3), 1853.

40. These articles appeared in the following issues: January 12 (2, 3), 14 (2, 4), 18 (2, 4), 22 (2, 4), 25 (2, 4), 27 (2, 4), 29 (2, 4–5), 1853.  41. *New-York Daily Tribune*, December 17 (4, 6), 1852.  42. *Rochester Daily American*, January 12 (2, 3–4), 1853.

43. *Ibid.*, January 29 (2, 5), 1853.  44. *Ibid.*, December 20 (2, 3), 1852.  45. *Ibid.*, January 29 (2, 4), 1853; January 7 (2, 3), 1850.  46. *New-York Daily Tribune*, December 11 (4, 6), 1852.  47. *Ibid.*, December 10 (4, 3), 1852.  48. *Rochester Daily American*, December 14 (2, 3), 1852.  49. *Genesee Farmer*, xiv (1853), 208.


CHAPTER VI

1. *Rochester Daily American*, June 1 (2, 3), 1850.  2. *Ibid.*, March 10 (2, 2), 1853.  3. *Ibid.*, December 31 (2, 3), 1851.  4. *Ibid.*, August 20 (2, 3), 1851.

5. For examples, *ibid.*, June 1 (2, 4), 1850; July 20 (2, 3), 1852; January 27 (2, 3), February 2 (2, 3), May 31 (2, 3), 1854.  6. *Ibid.*, January 31 (2, 3), 1854.  7. For examples, *ibid.*, January 29 (2, 3), February 20 (2, 2), February 22 (2, 3–4), April 5 (2, 3), April 19 (2, 3), June 6 (2, 3), June 19 (2, 3), June 27 (2, 3), 1850.

8. *Ibid.*, February 15 (2, 3), 1850.  9. *Ibid.*, June 19 (2, 3), 1850.  10. *Ibid.*, April 11 (2, 3), April 15 (2, 3), 1850.  11. *Ibid.*, July 16 (2, 3), July 17 (2, 3), 1850.  12. *Ibid.*, April 5 (2, 3), 1850.  13. *Ibid.*, April 20 (2, 3), June 25 (2, 3), 1850.  14. *Ibid.*, June 5 (2, 3), 1850.  15. *Ibid.*, January 28 (2, 3), 1850.  16. *Ibid.*, February 11 (2, 4), 1850.

17. *Ibid.*, February 13 (2, 5), March 14 (2, 2), 1850.

18. *Ibid.*, February 18 (2, 2), April 27 (2, 3), April 30 (2, 3), June 10 (2, 3), June 11 (2, 3), June 17 (2, 4), June 18 (2, 4), June 22 (2, 3), July 1 (2, 2), July 10 (2, 4), July 24 (2, 2), July 25 (2, 5), July 26 (2, 2), July 27 (2, 3), August 2 (2, 3), August 6 (2, 4), August 14 (2, 3), August 20 (2, 3), August 21 (2, 3), August 23 (2, 3), August 27 (2, 4), August 30 (2, 3), September 16 (2, 4), September 19 (2, 3), 1850.  19. *Ibid.*, June 17 (2, 3), 1850.

20. *Ibid.*, January 16 (2, 3), March 4 (2, 4), March 8 (2, 4), March 19 (2, 3), 1850.

21. *Ibid.*, May 30 (2, 3), 1854.  22. *Ibid.*, June 3 (2, 3), 1854.  23. *Ibid.*, March 26 (2, 3), 1853.  24. *Ibid.*, January 23 (2, 5), 1850; September 25 (2, 3), 1851.  25. *Ibid.*, June 21 (2, 3), 1854.  26. *Ibid.*, August 19 (2, 3), 1854. See chap. viii.  27. *Ibid.*, May 21 (2, 1), 1856.


CHAPTER VII

1. *Patent Office Report, 1849*, ii, 13.  2. *Plantation*, iii (May 8, 1872), 297.  3. *Country Gentleman*, xxx (February 23, 1865), 122.  4. *Southern Cultivator*, xlvii (July, 1889), 324.  5. *Genesee Farmer*, x (1849), 249.  6. *Patent Office Report, 1852*, ii, 16–22.  7. *Genesee Farmer*, xiii (1852), 41.  8. *Ibid.*, x (1849), 251.  9. *Ibid.*, xi (1850), 100.  10. Augusta *Weekly Constitutionalist*, April 13 (3, 1), 1859.

11. *Southern Cultivator*, xvi (January, 1858), 9; Albany (N.Y.) *Cultivator*, i (February, 1844), 47.  12. *Southern Cultivator*, xiv (April, 1856), 123.  13. *Genesee Farmer*, xii (1851), 274.  14. Augusta *Weekly Chronicle & Sentinel*, October 27 (2, 8), 1847.

15. *Southern Cultivator*, xi (July, 1853), 197.  16. *Genesee Farmer*, xiv (1853), 9.

17. *Patent Office Report, 1850*, ii, 42; *Southern Cultivator*, xviii (February, 1869), 48–49, (August, 1869), 240.  18. Albany (N.Y.) *Cultivator*, ii (August, 1845), 243.

19. *Patent Office Report, 1849*, ii, 304–5.  20. *Southern Cultivator*, xlvi (June, 1888), 257.

21. *Ibid.*, xlvii (February, 1889), 85.   22. Albany (N.Y.) *Cultivator*, ii (January, 1845), 35.   23. *Southern Cultivator*, xxiv (July, 1866), 161.   24. *Genesee Farmer*, x (1849), 11.   25. *Cultivator & Country Gentleman*, February 24, 1870, p. 115.
26. *Southern Cultivator*, xxvi (December, 1868), 357–58.   27. *Ibid.*, xxvii (August, 1869), 240.   28. *Ibid.*, xlvii (September, 1889), 433.   29. *American Farmer*, ii (April, 1868), 312–13.   30. *Ibid.*, x (December, 1852), 353–54. See also *ibid.*, xxv (October, 1867), 307.

31. *Ibid.*, xiv (June, 1856), 179. See also *ibid.*, xlvii (September, 1889), 433.   32. *Cultivator & Country Gentleman*, April 27, 1871, p. 260.   33. *Southern Cultivator*, xiii (July, 1855), 201–2.   34. *Southern Field and Fireside*, i (November 5, 1859), 190.

35. *Country Gentleman*, xi (September 18, 1858), 107.   36. *Plantation*, iii (June 26, 1872), 409.   37. *Patent Office Report, 1849*, ii, 7.   38. *Ibid., 1852*, ii, 4.   39. *Ibid.*, 1–15.   40. *Country Gentleman*, xxv (April 20, 1865), 250.

41. *Southern Cultivator*, xv (May, 1857), 138–39.   42. *Ibid.*, xiii (February, 1855), 41.   43. *Ibid.*, xv (August, 1857), 237.   44. Augusta *Weekly Chronicle & Sentinel*, January 31 (3, 3), 1849.   45. Athens *Southern Banner*, February 12 (2, 4–5), 1857.

46. *Plantation*, iii (June 26, 1872), 409.   47. *Farmer and Artisan*, i (January 4, 1870), 24.   48. *Patent Office Report, 1849*, ii, 13.   49. *Southern Cultivator*, xviii (January, 1855), 9.   50. *Ibid.*, xlvii (February, 1889), 85.

51. *Genesee Farmer*, xii (1851), 81.   52. *Ibid.*, x (1849), 277.   53. *Patent Office Report, 1850*, ii, 25–81.   54. Albany (N.Y.) *Cultivator*, iv (June, 1847), 33.   55. *Southern Cultivator*, xiv (January, 1856), 27–28. See also Athens *Southern Banner*, October 25 (3, 1–2), 1855.   56. *Southern Cultivator*, xiv (January, 1856), 17.   57. *Ibid.*, 27–28.

58. *Plantation*, iii (May 8, 1872), 296. See also *Southern Cultivator*, xxv (September, 1867), 272; *American Farmer*, i (January, 1867), 220–21.   59. *Southern Field and Fireside*, i (February 18, 1860), 310.   60. *Southern Cultivator*, xlvi (August, 1888), 355.

61. *Plantation*, iii (August 14, 1872), 514.   62. *Genesee Farmer*, vii (1846), 129.

63. *Ibid.*, xiii (1852), 206.   64. *Ibid.*, 37; *American Farmer*, iii (April, 1869), 308; *Southern Field and Fireside*, i (February 25, 1860), 318–19.

## CHAPTER VIII

1. *Southern Cultivator*, xv (May, 1857), 138.   2. *Ibid.*, 138–39.   3. *Ibid.*, xlvii (September, 1889), 437.   4. *Patent Office Report, 1849*, ii, 12; *Plantation*, iii (September 11, 1872), 585.   5. *Southern Cultivator*, xlviii (February, 1890), 77.   6. *Ibid.*, (March, 1890), 112.   7. *Ibid.*, xxvi (January, 1868), 22.   8. *Genesee Farmer*, xiv (1853), 271.

9. *Plantation*, iii (May 8, 1872), 297.   10. *Southern Cultivator*, xlvii (September, 1889), 437.

11. *Ibid.*, (October, 1889), 501.   12. *Ibid.*, xv (October, 1857), 314. See also *ibid.*, xv (December, 1857), 362.   13. Augusta *Weekly Constitutionalist*, January 21 (5, 3), 1857. See also *Southern Cultivator*, xv (November, 1857), 346.   14. *Southern Cultivator*, xiv (October, 1856), 298–300; xv (April, 1857), 120; Bonner, *Georgia Agriculture*, 83.   15. Augusta *Weekly Constitutionalist*, September 9 (4, 1), 1857.   16. *Country Gentleman*, xi (April 1, 1858), 202.   17. *American Cotton Planter and the Soil of the South . . .*, i (August, 1857), 243.   18. Athens *Southern Watchman*, May 6 (2, 6; 3, 1), 1863.

19. *Southern Cultivator*, xv (November, 1857), 341.   20. *Patent Office Report, 1849*, ii, 231–36; *Cultivator & Country Gentleman*, xxxii (November 24, 1870), 739.

21. *Patent Office Report, 1849*, ii, 207.   22. *Plantation*, iii (September 11, 1872), 585.   23. *Southern Cultivator*, xiii (October, 1855), 297; xxv (October, 1867), 309.   24. *Ibid.*, xlvii (January, 1889), 17.   25. *Plantation*, iii (August 21, 1872), 529.   26. *Southern Cultivator*, xii (June, 1854), 170.   27. *Ibid.*, xxix (March, 1871), 85.   28. *Southern Field and Fireside*, iii (August 31, 1861), 119.   29. *Southern Cultivator*, xxvii (May, 1869), 172.

30. *Genesee Farmer*, xiv (1853), 201.   31. *Ibid.*, x (1849), 12.   32. *Ibid.*, xiii (1852),

333. 33. *Southern Cultivator*, xv (March, 1857), 91–92. See also *Plantation*, iii (August 14, 1872), 521. 34. *Country Gentleman*, xxv (April 14, 1865), 233. 35. *Cultivator & Country Gentleman*, xxxii (November 17, 1870), 723, (November 24, 1870), 739; (December 1, 1870), 755. See also *ibid.*, xxix (February 21, 1867), 122. 36. *Ibid.*, xxxii (December 8, 1870), 771. 37. *Ibid.*, xxix (April 25, 1867), 283. 38. *Southern Cultivator*, xxv (December, 1867), 565. 39. *Ibid.*, xxv (March, 1867), 69. 40. *Ibid.*, xlv (July, 1887), 292.

41. *American Farmer*, iv (July, 1869), 5–6; *Southern Cultivator*, xxvii (July, 1869), 208; xlv (October, 1887), 458; xlvii (September, 1888), 411. 42. *Southern Field and Fireside*, iii (May 11, 1861), 406. 43. *Southern Cultivator*, xxvii (September, 1869), 276. 44. *Plantation*, iii (July 24, 1872), 473. 45. *Southern Field and Fireside*, ii (May 26, 1860), 6. 46. *Southern Cultivator*, xxvii (May, 1869), 172. 47. *Ibid.*, xiv (March, 1856), 90. 48. *Southern Field and Fireside*, iii (April 12, 1862), 292; *Southern Cultivator*, xxv (October, 1867), 310; xxvii (July, 1869), 222.

49. *Southern Cultivator*, xxvii (August, 1869), 247. 50. *Ibid.*, xiii (October, 1855), 316. 51. *Ibid.*, xlvi (August, 1888), 353. 52. *Ibid.*, xlvi (April, 1888), 161. 53. *Plantation*, iii (July 24, 1872), 473. 54. *Southern Cultivator*, xlvi (February, 1888), 68. See also *ibid.* (April, 1888), 161. 55. *Cultivator & Country Gentleman*, xxxii (September 22, 1870), 595. 56. *Farmer and Artisan*, i (January 4, 1870), 25. 57. *Cultivator & Country Gentleman*, xxxiii (January 26, 1871), 51. 58. *Southern Field and Fireside*, iii (May 11, 1861), 406. 59. *Southern Cultivator*, xxiv (December, 1866), 280.

60. *American Farmer*, iv (December, 1869), 174–76.

61. *Cultivator & Country Gentleman*, xxxii (September 22, 1870), 595. 62. *Southern Cultivator*, xlvii (September, 1888), 411. 63. *Cultivator & Country Gentleman*, xxxiii (January 26, 1871), 51. 64. *Southern Field and Fireside*, iii (December 14, 1861), 244. 65. *Plantation*, iii (July 31, 1872), 487. 66. *Cultivator & Country Gentleman*, xxxiii (March 9, 1871), 149. 67. *Southern Field and Fireside*, iii (February 15, 1862), 260. 68. *Ibid.*, iii (July 13, 1861), 62. 69. *Ibid.*, iii (November 16, 1861), 207; *Southern Cultivator*, xx (February, 1862), 49. 70. *Southern Field and Fireside*, iii (July 13, 1861), 62.

71. *Plantation*, iii (July 31, 1872), 487; *Southern Cultivator*, xi (January, 1853), 25. 72. *American Agriculture*, iii (July, 1844), 211–12; *Country Gentleman*, xxv (March 16, 1865), 172; Albany (N.Y.) *Cultivator*, i (February, 1844), 47; (March, 1844), 94. 73. *Country Gentleman*, xxv (January 12, 1865), 28. 74. *Cultivator & Country Gentleman*, xxxiii (November 30, 1871), 761. 75. *Genesee Farmer*, ix (1848), 286. 76. *Ibid.*, x (1849), 106; xiii (1852), 143–46. 77. *Plantation*, iii (May 15, 1872), 315. 78. *Southern Cultivator*, xxvii (November, 1869), 340. 79. *Ibid.*, xiii (March, 1855), 74. 80. *Ibid.*, xlv (June, 1887), 341.

81. *Plantation*, iii (December, 1872), 88. 82. Augusta *Weekly Chronicle & Sentinel*, November 24 (3, 5), 1847. 83. *Genesee Farmer*, ix (1848), 37. 84. *Ibid.*, 216.

85. *Southern Cultivator*, xxv (June, 1867), 163–64. 86. *American Farmer*, i (April, 1867), 301–2. 87. *Southern Field and Fireside*, iv (June 7, 1862), 12.


## CHAPTER IX

1. For a discussion of this subject, see McManus, *Negro Slavery in New York*, passim. 2. *Southern Cultivator*, xv (April, 1857), 107. 3. *Ibid.*, xii (August, 1854), 234. 4. Augusta *Weekly Chronicle & Sentinel*, October 10 (1, 2), 1849. 5. *Genesee Farmer*, xi (1850), 201. 6. *Southern Cultivator*, xv (March, 1857), 91. 7. Augusta *Weekly Constitutionalist*, March 23 (3, 1), 1859. 8. *Ibid.*, March 9 (3, 1), 1859. 9. *Genesee Farmer*, ix (1848), 37. 10. *Southern Field and Fireside*, i (December 31, 1859), 254.

11. *Southern Cultivator*, xv (March, 1857), 76. 12. *Genesee Farmer*, x (1849), 82. 13. *Acts of the General Assembly of the State of Georgia, Passed in Milledgeville, at the Annual Session in November and December, 1859* (Milledgeville, Ga., 1860),

69–70. 14. *Southern Field and Fireside*, I (January 14, 1860), 270; (January 28), 286; (February 4), 294; (May 5), 399; (May 12), 406. 15. *Southern Cultivator*, XIV (July, 1856), 202–4. 16. *Ibid.*, XII (June, 1854), 169. 17. *Ibid.*, XV (March, 1857), 75. 18. *Ibid.*, (May, 1857), 147. See also *Patent Office Report, 1849*, II, 307–13. 19. W. S. Jenkins, *Pro-Slavery Thought in the Old South* (Chapel Hill, 1935), 95–103.
  20. Augusta *Weekly Constitutionalist*, February 16 (2, 2), 1859. 21. *Ibid.*, March 23 (3, 1), 1859. 22. *Ibid.*, March 16 (3, 1), 1859. 23. *Ibid.*, March 9 (4, 3), 1859. 24. *Ibid.*, February 16 (2, 2), 1859. 25. *Ibid.*, February 16 (2, 3), 1859. 26. *Ibid.*, March 30 (3, 1), 1859. 27. *Ibid.*, March 2 (4, 2), 1859. 28. *Ibid.*, March 16 (1, 3), 1859. 29. *Ibid.*, March 2 (3, 1), 1859. 30. *Ibid.* 31. *Ibid.*, February 16 (1, 4), 1859. 32. *Ibid.*, February 23 (1, 6), 1859. 33. *Ibid.*, March 16 (3, 1), 1859. 34. *Ibid.*, March 30 (1, 3), 1859.
  35. *Southern Cultivator*, XVI (August, 1858), 235. 36. Augusta *Weekly Constitutionalist*, March 2 (3, 1), 1859. 37. *Ibid.* 38. *Ibid.*, February 23 (1, 5–6), 1859. 39. *American Cotton Planter and Soil of the South*, II (October, 1858), 325. 40. *Southern Cultivator*, XVII (March, 1859), 84. 41. Augusta *Weekly Constitutionalist*, April 20 (4, 2–3), 1859. 42. *Southern Cultivator*, XVI (May, 1858), 138. 43. *Genesee Farmer*, X (1849), 42. 44. *Rochester Daily American*, June 12 (2, 3), 1854. 45. *Southern Cultivator*, XII (April, 1854), 107. 46. *Ibid.* (June 1854), 169. 47. *Rochester Daily Democrat*, June 6 (2, 2), 1854. 48. *Ibid.*
  49. *Rochester Daily American*, June 13 (2, 3), 1854. 50. *Ibid.*, June 12 (2, 3), 1854. 51. *Ibid.*, June 7 (2, 2), 1854. 52. *Rochester Daily Democrat*, June 6 (2, 2), 1854. 53. *Ibid.*, *Southern Cultivator*, XII (May, 1854), 170. 54. *Rochester Daily American*, June 15 (2, 3), 1854. 55. *Ibid.*, June 21 (2, 3), 1854. 56. *Ibid.*, June 9 (2, 2), 1854. Copied in *Rochester Daily Democrat*, June 12 (2, 3), 1854. 57. *Rochester Daily Democrat*, June 12 (2, 1; 2, 3), 1854. 58. *Southern Cultivator*, XII (June, 1854), 170. 59. *Rochester Daily American*, June 14 (2, 2), 1854.
  60. *Ibid.*, XII (April, 1854), 106. 61. *Ibid.*, 107. 62. *Rochester Daily American*, June 20 (2, 3), 1854. 63. *Ibid.*, June 29 (2, 3), 1854. 64. *Southern Field and Fireside*, I (December 31, 1859), 254. 65. *Ibid.*, I (November 24, 1859), 214. 66. *Southern Cultivator*, XIV (September, 1856), 282–83. See also *ibid.* (July, 1856), 205–6. 67. *Genesee Farmer*, IX (1848), 9. 68. *Ibid.*, XII (1851), 36. 69. *Ibid.*, IX (1848), 9. 70. *Journal of the United States Agricultural Society*, I, 27. 71. *Genesee Farmer*, XI (1850), 9.
  72. *Ibid.*, IX (1848), 120.

## CHAPTER X

1. Eliza A. Bowen, *The Story of Wilkes County, Georgia*, reprint ed. (Marietta, Ga., 1950), 94; *Biographical Directory of the American Congress, 1774–1927* (Washington, D.C., 1928), 1604; Augusta *Weekly Chronicle & Sentinel*, July 11 (3, 1), 1855; Savannah *Courier*, n.d., quoted in *Southern Cultivator*, XIII (August, 1855), 250.
  2. Hancock County Record Book, T (Sparta, Ga.), 6–31; Augusta *Weekly Chronicle & Sentinel*, December 10 (3, 2), 1854; *Southern Cultivator*, XII (June, 1854), 188; XIII (February, 1855), 46.
  3. Augusta *Weekly Chronicle & Sentinel*, December 10 (3, 2), 1854.
  4. *Southern Cultivator*, XIII (August, 1855), 250; Athens *Southern Banner*, July 12 (2, 7), 1855; Augusta *Weekly Chronicle & Sentinel*, July 11 (3, 1), 1855.
  5. Minutes of the Board of Trustees of the University of Georgia, 1835–1857, III, 264 (typescript), 333 (manuscript). 6. *Genesee Farmer*, IX (1848), 9. 7. *Southern Cultivator*, IV (October, 1846), 153. 8. *Genesee Farmer*, XII (1851), 289. 9. *Southern Field and Fireside*, III (July 20, 1861), 70. 10. *Endowment of the Terrell Professorship of Agriculture*, 3–4. This information is included also in the University Trustee Minutes, III, 241–44 (typescript), 303–9 (manuscript).
  11. *Endowment of the Terrell Professorship of Agriculture*, 5–6.
  12. *Ibid.*, 8. 13. *Ibid.*, 9–10. 14. *Ibid.*, 10, 11. 15. *Ibid.*, 11,12. 16. *Georgia Uni-*

*versity Magazine*, October, 1854, 287. 17. *Ibid.*, February, 1855, 350. See also *ibid.*, October, 1854, 287. 18. Augusta *Weekly Chronicle & Sentinel*, August 16 (2, 6), 1854. 19. *Southern Cultivator*, xiii (August, 1855), 250, quoting Savannah *Courier*, n.d. 20. *Southern Cultivator*, xiii (January, 1855), 18, quoting Savannah *Courier*, n.d. 21. *Southern Cultivator*, xiii (April, 1855), 115. 22. *Ibid.* (July, 1855), 214. See also *ibid.* (May, 1855), 159. 23. *Genesee Farmer*, xv (1854), 289. 24. *Southern Cultivator*, xii (September, 1854), 281. 25. *Endowment of the Terrell Professorship of Agriculture*, 15.

## CHAPTER XI

1. *Endowment of the Terrell Professorship of Agriculture*, 6, 7; University Faculty Minutes, 1850–1873, 40.
2. *Report of President Church of the State University, to the Senatus Academicus, Held in the Senate Chamber, Thursday, November 8th, 1855* (Milledgeville Ga., n.d.), 3; Athens *Southern Banner*, October 4 (2, 6), 1855; *Southern Cultivator*, xvi (May, 1858), 145. 3. *Ibid.*, xiii (April, 1855), 116; *The Soil of the South . . .* , v (May, 1855), 131.
4. *University of Georgia. Catalogue of the Officers and Students of Franklin College, Athens, Geo., 1854–'55* (Athens, Ga., n.d.), 17. 5. *Southern Cultivator*, xiii (May, 1855), 138–40. 6. *Ibid.*, 137; *Soil of the South*, v (May, 1855), 137–43.
7. Quoted in *Southern Cultivator*, xiii (May, 1855), 159. 8. *Soil of the South*, v (May, 1855), 131. 9. University Trustee Minutes, 1835–1857, iii, 272 (typescript), 343 (manuscript). 10. *Southern Cultivator*, xv (February, 1857), 42–45, 74–76; Athens *Southern Banner*, February 12 (1, 2–7), 1857.
11. Athens *Southern Banner*, October 11 (2, 7), 1855. 12. *Southern Cultivator*, xvii (March, 1859), 66–68, 98–100. 13. *Ibid.*, xvi (May, 1858), 145. 14. *Ibid.*, xvii (September, 1859), 259.
15. *Ibid.*, xv (August, 1857), 237. 16. *Country Gentleman*, xxv (February 9, 1865), 91. 17. *Southern Field and Fireside*, ii (December 8, 1860), 230. 18. *Southern Cultivator*, xiii (January, 1855), 9. 19. *Ibid.*, xv (September, 1857), 279. 20. *Endowment of the Terrell Professorship of Agriculture*, 11.
21. University Trustee Minutes, 1835–1857, iii, 262–63 (typescript), 331 (manuscript). 22. Augusta *Weekly Constitutionalist*, December 9 (8, 3), 1857. 23. *Southern Cultivator*, xiv (June, 1887), 250. 24. *Ibid.*, xiii (January, 1855), 18, 19. See also *Southern Field and Fireside*, ii (March 30, 1861), 358. 25. *Southern Cultivator*, xiii (January 1855), 9–10.
26. Augusta *Weekly Constitutionalist*, December 9 (8, 3), 1857. In his report to the University trustees, Lee asked for $100 for purchasing books on agriculture.
27. *Southern Cultivator*, xvi (May, 1858), 145; xxvi (December, 1868), 358; Athens *Southern Banner*, July 24 (3, 4), 1856. 28. Augusta *Weekly Chronicle & Sentinel*, August 9 (2, 8), 1854; Augusta *Weekly Constitutionalist*, December 9 (8, 1–3), 1857; *Report of President Church*, 3. 29. Augusta *Weekly Constitutionalist*, December 9 (8, 1), 1857.
30. University Trustee Minutes, 1835–1857, iii, 279–305 (typescript), 352–81 (manuscript); Athens *Southern Banner*, October 11 (2, 7), 1855. 31. *Journal of the Senate of the State of Georgia, at the Biennial Session . . . 1855 & 1856* (Milledgeville, Ga., 1855 [sic]), 29. 32. Athens *Southern Banner*, February 14 (2, 7), 1856. 33. *Journal of the Senate of Georgia, 1855 & 1856*, 493–95. 34. Augusta *Weekly Constitutionalist*, December 9 (8, 1), 1857. 35. University Trustee Minutes, 1835–1857, iii, 317 (typescript), 395 (manuscript); Augusta *Weekly Constitutionalist*, December 9 (8, 3), 1857.
36. For instance, *Catalogue of the University of Georgia, 1855–1856*, 20. 37. University Trustee Minutes, 1835–1857, iii, 259, 289, 345–46 (typescript), 328, 363, 432–33 (manuscript). 38. *Ibid.*, iii, 249 (typescript), 315 (manuscript); *ibid.*, 1857–1871, iv,

25–26 (typescript), 35 (manuscript). 39. See E. M. Coulter, "Why John and Joseph LeConte Left the University of Georgia, 1855–1856," in *Georgia Historical Quarterly*, LIII (March, 1969), 18–40.

40. *Southern Field and Fireside*, I (December 3, 1859), 222. 41. *Ibid.*, I (October 29, 1859), 182. 42. University Trustee Minutes, 1857–1871, IV, 43–44, 55 (typescript), 61–62, 77–78 (manuscript); *Southern Field and Fireside*, II (August 18, 1860), 103.

43. University Trustee Minutes, 1857–1871, IV, 54 (typescript), 76 (manuscript).

44. *Southern Field and Fireside*, III (July 20, 1861), 70. 45. University Trustee Minutes, 1857–1871, IV, 52, 53 (typescript), 73, 75 (manuscript); Augusta *Weekly Constitutionalist*, July 17 (4, 3), 1861.

46. University Trustee Minutes, 1857–1871, IV, 65–66 (typescript), 92–93 (manuscript). 47. *Ibid.*, IV, 84 (typescript), 119 (manuscript); University Faculty Minutes, 1850–1873, pp. 40–170; Athens *Southern Watchman*, September 9 (2, 1), 1863.

48. *Southern Field and Fireside*, n.s. I (November 15, 1863). 49. *Cultivator & Country Gentleman*, XXX (December 26, 1867), 410. 50. Clarke County (Ga.) Deed Record, V, 190.

51. Eighth Census, 1860, Georgia, Appling-Fayette, Agriculture (MS microfilm), 27–28. See also *Southern Field and Fireside*, II (July 27, 1861), 79; *Southern Cultivator*, XIII (March, 1855), 90. 52. Eighth Census, 1860, Georgia, Appling-Fayette, Agriculture (MS microfilm), 27–28. 53. *Southern Field and Fireside*, I (August 27, 1859), 110. 54. *Ibid.*, II (October 20, 1860), 174. 55. Clarke County (Ga.) Deed Record, V, 260, 263; Eighth Census, 1860, Georgia, Cass-DeKalb, Population (MS microcopy T–7, roll 27), 30, 130; *Southern Field and Fireside*, I (August 13, 1859), 94; Augusta *Weekly Constitutionalist*, February 16 (2, 2), 1859; *Plantation*, III (August 28, 1872), 551. Regarding Lee's age, see n. 2, chap. I.

56. *Genesee Farmer*, XV (1854), 226. 57. *Cultivator & Country Gentleman*, June 1, 1871, p. 345; *Southern Cultivator*, XLVI (June, 1888), 257; Athens *Southern Watchman*, May 10 (3, 1–2), 1860. 58. *Southern Cultivator*, XLVI (April, 1888), 157. 59. Athens *Southern Watchman*, May (1, 3–4), 1860. 60. *Ibid.*, April 12 (3, 1), 1860.

61. *Southern Field and Fireside*, III (July 6, 1861), 54. 62. *Ibid.*, III (March 8, 1862), 272. 63. *Ibid.*, I (August 27, 1859), 110; III (November 16, 1861), 207; Athens *Southern Watchman*, May 10 (3, 1–2), 1860. 64. *Southern Field and Fireside*, I (August 27, 1859), 110. 65. *Plantation*, III (December, 1872), 68; *Southern Cultivator*, XXVII (March, 1869), 81. 66. *Southern Field and Fireside*, II (June 15, 1861), 30. 67. *Tennessee Farmer*, April 14, 1887, 4. 68. *Southern Cultivator*, XXV (September, 1867), 281. 69. *Plantation*, II (June 10, 1871), 313. 70. *Ibid.*, II (July 1, 1871), 353.

71. *Ibid.*, III (May 22, 1872), 329. 72. *Southern Cultivator*, XXXIV (November, 1876), 458.

## CHAPTER XII

1. *Southern Field and Fireside*, I (December 10, 1859), 3. 2. Augusta *Weekly Constitutionalist*, March 30 (5, 1–2), 1859. 3. *Southern Field and Fireside*, I (May 28, 1859), 5. 4. *Ibid.*, IV (November 15, 1862), 99. When the journal resumed publication on January 3, 1863, V. Lataste was listed as editor of the agricultural department, without any mention of Lee.

5. Augusta *Weekly Chronicle & Sentinel*, May 9 (1, 5), 1849. 6. *Southern Cultivator*, XI (March, 1853), 72. 7. *Southern Field and Fireside*, I (August 27, 1859), 110.

8. *Southern Cultivator*, XIX (March, 1861), 100. 9. *Ibid.*, XV (May, 1857), 147. 10. *Ibid.*, X (February, 1852), 39.

11. *Southern Field and Fireside*, III (October 5, 1861), 159. 12. *Ibid.*, II (December 1, 1860), 222; II (January 5, 1861), 262. 13. *Ibid.*, II (January 5, 1861), 262; II (March 16, 1861), 311. 14. *Ibid.*, II (December 15, 1860), 238. 15. *Ibid.*, II (January 19, 1861), 279. 16. *Ibid.* 17. *Ibid.*, II (January 1, 1861), 11. 18. *Ibid.*, III (July 27, 1861), 79. 19. *Ibid.*, II (April 27, 1861), 390. 20. *Ibid.*, III (January 25, 1862), 248. 21. *Ibid.*, III

(April 5, 1862), 288.  22. *Ibid.*, III (January 18, 1862), 244.  23. *Ibid.*, II (February 23, 1861), 318; III (August 21, 1861), 111.  24. *Ibid.*, III (March 15, 1862), 276.  25. *Ibid.*, III (February 8, 1862), 256; III (May 11, 1861), 406.  26. *Ibid.*, II (August 3, 1861), 87. 27. *Ibid.*, II (August 17, 1861), 103.

28. Athens *Southern Watchman*, January 1 (1, 2–4); December 3 (2, 6), 1862. 29. *Southern Field and Fireside*, III (May 25, 1861), 6.  30. *Ibid.*, II (March 23, 1861), 350.  31. Athens *Southern Watchman*, December 24 (3, 3), 1862.  32. *Southern Field and Fireside*, III (November 9, 1861), 199.  33. *Ibid.*, II (March 16, 1861), 342.  34. *Ibid.*, IV (June 28, 1862), 24; August 16, p. 48; November 1, p. 92; November 15, p. 100.

35. *Plantation*, III (September 11, 1872), 583; *Farmer and Artisan*, I (January 4, 1870), 26; *Southern Field and Fireside*, IV (June 21, 1862), 20; *ibid.* (July 5, 1862), 28. 36. Knox County (Knoxville, Tenn.) Deed Record, B3, 193, 216; D3, 51. For another purchase in Knoxville, May 14, 1863, see *ibid.*, B3, 354.  37. *Cultivator & Country Gentleman*, XXX (December 26, 1867), 410. No record of this transaction was found in the deed records of Knox County in the courthouse in Knoxville, Tennessee.  38. And, then, much of anything that had been saved from Burnside's troops probably went up in smoke later, for again he lost his house—this time through the negligence of a servant. *Plantation*, III (July 31, 1872), 487.  39. *Cultivator & Country Gentleman*, XXX (July 25, 1867), 59.  40. *Ibid.*, XXX (December 26, 1867), 410.  41. *Country Gentleman*, XXV (April 20, 1865), 250.

## CHAPTER XIII

1. Rochester *Union & Advertiser*, September 7 (2, 2), 1865; *New York Times*, September 10 (5, 3), 1865.

2. Knox County (Knoxville, Tenn.) Deed Record, B3, 532–33.  3. *Ibid.*, D3, 158. See also 411.  4. *Ibid.*, L3, 126; Q3, 437.  5. *Ibid.*, L3, 517.  6. Knox County (Knoxville, Tenn.) Trust Deed, B, I, 614.  7. Trust Deed, R3, 253.  8. *Ibid.*, D3, 542; F3, 527.  9. *Ibid.*, H3, 475; N3, 62.  10. *Ibid.*, D3, 64, 410.  11. *Ibid.*, L3, 580; M3, 158; R3, 190.  12. *Ibid.*, D3, 62; Clarke County (Ga.) Deed Record, X, 281.

13. *Cultivator & Country Gentleman*, XXX (July 25, 1867), 59; XXXIII (April 27, 1871), 260; *Southern Cultivator*, XLVI (September, 1888), 409.  14. *Southern Cultivator*, XXV (August, 1867), 250.  15. *Ibid.*, XXVII (July, 1869), 222.  16. *Cultivator & Country Gentleman*, XXIX (January 3, 1867), 14.  17. *Ibid.*  18. *Southern Cultivator*, XLVI (September, 1888), 409.

19. *American Farmer*, II (January, 1868), 214; *Plantation*, III (August 14, 1872), 414. 20. *Southern Cultivator*, XXIV (December, 1866), 280.  21. *Ibid.*, XXV (August, 1867), 250; *American Farmer*, I (May, 1867), 338–39.  22. *Southern Cultivator*, XLVI (February, 1888), 68.  23. *Ibid.*, XLVII (September, 1889), 448.  24. *Ibid.*, XLV (July, 1887), 292.  25. *Plantation*, III (May 29, 1872), 346.

26. *Southern Cultivator*, XXVI (May, 1868), 131.  27. *Ibid.* (August, 1868), 228. 28. *Plantation*, III (July 17, 1872), 455.

29. *Southern Cultivator*, XXV (July, 1867), 197.  30. *Cultivator & Country Gentleman*, XXX (July 25, 1867), 59.  31. *Genesee Farmer*, XIII (1852), 137.  32. *Southern Cultivator*, XXVII (November, 1869), 340.  33. *Ibid.*, XXVII (May, 1869), 172.  34. *Ibid.* 35. *Plantation*, III (September 25, 1872), 609.

36. *Cultivator & Country Gentleman*, XXXIII (June 1, 1871), 345.  37. *Plantation*, III (October 2, 1872), 633.  38. *Southern Cultivator*, XLVI (February, 1888), 68.  39. *Plantation*, III (May 15, 1872), 312.  40. *Ibid.* (August 7, 1872), 496.  41. *Southern Cultivator*, XXVI (October, 1868), 292.

42. *Cultivator & Country Gentleman*, XXIX (June 3, 1867), 14. Lee was deeply upset by conditions in the South where minority governments were in control, but he was especially worried by Brownlow's regime. Not able to keep his silence longer from his Rochester friends, he wrote one of them a letter in 1867 describing the revolution then in progress. The Rochester *Union & Advertiser* (April 26 [2, 3], 1867) published

it with this comment by the editor: "Those who know the author will give him credit for candor, and they will admit that he is a man likely to be pretty well informed as to public sentiment in his State at least."

Pouring out his heart, Lee said, "A small minority of the adult male population of Tennessee has disfranchised the majority; and to perpetuate its power in the State against the consent of the majority, it is now organizing for active service all over Tennessee, companies, battalions and regiments of armed men. Is this peace or war? Can farmers have their horses and corn stolen, their meat houses and poultry houses robbed, after the well-known custom of soldiers and camp followers, and not feel deeply the weight of an iron yoke placed on their necks for the imputed crime of political heresy?"

43. *Cultivator & Country Gentleman*, xxx (December 26, 1867), 411. See also 410.
44. See E. Merton Coulter, *William G. Brownlow: Fighting Parson of the Southern Highlands* (Chapel Hill, 1937), 346–48. 45. *Southern Cultivator*, xi (May, 1853), 142. 46. *Cultivator & Country Gentleman*, xxiv (April 4, 1864), 221. 47. *Ibid.*, xxx (November 14, 1867), 316; *American Farmer*, ii (January, 1868), 208. 48. *Cultivator & Country Gentleman*, xxix (April 4, 1867). 49. *Ibid.* (January 3, 1867), 14–15. 50. *Ibid.*, 14.

51. *Southern Cultivator*, xlvii (January, 1889), 21. 52. *Ibid.*, xxv (March, 1867), 69. 53. *Ibid.*, xxvi (March, 1868), 77. 54. *Farmer and Artisan*, i (January 4, 1870), 25. See also 24, 26. 55. *Plantation*, iii (May 8, 1872), 296; (August 21, 1872), 535. 56. *Ibid.* (May 22, 1872), 330. See also October 2, 1827, 634; December, 1827, 91.

57. *Tennessee Farmer*, February 24, 1887, 1; *Southern Cultivator*, xlv (September, 1887), 400. 58. *Tennessee Farmer*, March 10, 1887, 2.

59. *Southern Cultivator*, xlvi (February, 1888), 67. See also xlv (June, 1887), 248; xlvii (February, 1889), 102. 60. *Ibid.*, xlvii (February, 1889), 102. 61. *Ibid.*, xlvi (May, 1888), 212, quoting *Weekly Star*. 62. *Tennessee Farmer*, April 7, 1887, 4.

63. *Southern Cultivator*, xlvii (September, 1889), 433. 64. *Plantation*, iii (August 14, 1872), 519. 65. *Ibid.* (August 21, 1872), 535. 66. *Ibid.* (December, 1872), 90.

67. *Southern Cultivator*, xlvii (January, 1889), 21. For other Lee writings (often repeating what he had previously written about), see *Plantation*, iii (June 12, 1872), 375–76; *Southern Cultivator*, xlvi (August, 1888), 353; *Tennessee Farmer*, March 17, 1887, 1, 4; April 7, 4; April 21, 2; May 5, 5; May 12, 4; May 19, 4; May 26, 1; June 2, 5; August 11, 4; October 13, 4.

68. Knox County Deed Record, J3, 530. 69. *Ibid.*, no. 112, 409. 70. Knox County Trust Deeds, J1, 644. 71. Knox County Deed Record, G4, 546. 72. *Southern Cultivator*, xlvi (April, 1888), 157. 73. *Ibid.* (November, 1888), 515. For an interesting description of Belle Meade, see William T. Alderson and Robert M. McBride, eds., *Landmarks of Tennessee History* (Nashville, 1965), 25–44. 74. *Southern Cultivator*, xlvii (July, 1889), 325. 75. Davidson County (Tenn.) Deed Record, no. 51, 207. See also no. 58, 570; *Southern Cultivator*, xlvii (February, 1889), 85.

76. Davidson County Deed Record, no. 72, 525; no. 98, 311; no. 131, 101. 77. *Southern Cultivator*, xlvi (September, 1888), 410, quoting the Atlanta *Capitol*, n.d.

78. *Southern Cultivator*, xlvi (October, 1888), 476. 79. *Atlanta Journal*, March 17 (6, 1), 1890. See also *Southern Cultivator*, xlvii (January, 1889), 4. 80. Joseph Buckner Killebrew Papers, "Reminiscences," I, 251, in Manuscript Section, Tennessee State Library and Archives, Nashville. This item was kindly furnished by Harriet C. Owsley (Mrs. Frank L. Owsley), director. Letter, Nashville, December 1, 1968.

81. *Southern Cultivator*, xlvii (October, 1889), 501. 82. *Southern Cultivator*, xlviii (April, 1890), 173; *Atlanta Journal*, March 17 (6, 1), 1890. 83. Davidson County (Tenn.) Wills and Inventories, book 30, 436; book 31, 522; Wills, book 33, 163.

84. This writer searched diligently through the files of the *Banner* but found no mention of Lee's death.

# Bibliography

## I. BOOKS

Alderson, William T., and Robert M. McBride, eds. *Landmarks of Tennessee History.* Nashville: Tennessee Historical Society and Tennessee Historical Commission, 1965.

Bailey, L. H. *Cyclopedia of American Agriculture.* 4 vols. New York: Macmillan, 1909.

*Biographical Directory of the American Congress, 1774–1927.* Washington, D.C.: Government Printing Office, 1928.

Bonner, James C. *A History of Georgia Agriculture, 1732–1860.* Athens, Ga.: University of Georgia Press, 1964.

Bowen, Eliza A. *The Story of Wilkes County, Georgia.* Reprint ed. Marietta, Ga.: Continental Book, 1950.

Coulter, E. Merton. *William G. Brownlow: Fighting Parson of the Southern Highlands.* Chapel Hill: University of North Carolina Press, 1937.

Demaree, Albert Lowther. *The American Agricultural Press, 1819–1860.* New York: Columbia University Press, 1941.

Gates, Paul W. *The Farmer's Age: Agriculture, 1815–1860.* Vol. III, *The Economic History of the United States.* New York: Holt, Rinehart & Winston, 1960.

Jenkins, W. S. *Pro-Slavery Thought in the Old South.* Chapel Hill: University of North Carolina Press, 1935.

Lewis, David W. *Transactions of the Southern Central Agricultural Society, from its Organization in 1846 to 1851, with an Introduction, Giving the Origin and Brief History of the Society.* Macon, Ga.: Benjamin F. Griffin, 1852.

McManus, Edgar J. *A History of Negro Slavery in New York.* Syracuse, N.Y.: Syracuse University Press, 1966.

True, Alfred Charles. *A History of Agricultural Education in the United States, 1785–1925.* United States Department of Agriculture Miscellaneous Publications, no. 36. Washington, D.C.: Government Printing Office, 1929.

## II. GOVERNMENT DOCUMENTS (Printed)

### A. Georgia

*Acts of the General Assembly of the State of Georgia, Passed in Milledgeville, at the Annual Session in November and December, 1859.* Milledgeville, Ga.: Boughton, Nisbet & Barnes, 1860.

*Journal of the Senate of the State of Georgia, at the Biennial Session . . . 1855 & 1856.* Milledgeville, Ga.: Boughton, Nisbet & Barnes, 1855 [*sic*].

### B. United States

*Report of the Commissioner of Patents for the Year 1845.* House Document no. 140, 29 Cong., 1 Sess. Washington, D.C.: Ritchie & Heiss, [1846].

*Ibid., 1849.* Part II, Agriculture. House Executive Document no. 20, 31 Cong., 1 Sess. Washington, D.C.: Office of Printers to the House of Representatives, 1850.

*Ibid., 1850.* Part II, Agriculture. House Executive Document no. 32, 31 Cong., 2 Sess. Washington, D.C.: Office of Printers to the House of Representatives, 1851.

*Ibid., 1851.* Part II, Agriculture. House Executive Document no. 102, 32 Cong., 1 Sess. Washington, D.C.: Robert Armstrong, 1852.

*Ibid., 1852.* Part II, Agriculture. Senate Executive Document no. 55, 32 Cong., 2 Sess. Washington, D.C.: Robert Armstrong, 1853.

*The Yearbook of Agriculture, 1940: Farmers in a Changing World.* Washington, D.C.: Government Printing Office, n.d.

## III. MANUSCRIPTS (Official and Unofficial)

Clarke County (Ga.) Deed Record, nos. v, w, z. County courthouse, Athens.

Davidson County (Tenn.) Deed Record, nos. 51, 58, 72, 98, 131. County courthouse, Nashville.

Davidson County (Tenn.) Wills and Inventories, nos. 30, 31, 33. County courthouse, Nashville.

Thomas Ewing Family Papers. Box 55. Manuscripts Division, Library of Congress.

Hancock County (Ga.) Records 1812–1858. Indexes in office of the clerk and office of the ordinary. County courthouse, Sparta.

Joseph Buckner Killebrew Papers, "Reminiscences." Manuscript Section. Tennessee State Library and Archives, Nashville.

Knox County (Tenn.) Deed Record, nos. B3, D3, F3, G3, H3, J3, L3, N3, R3, X, 112. County courthouse, Knoxville.
Knox County (Tenn.) Trust Deeds, no. J1. County courthouse, Knoxville.
Minutes of the Faculty, University of Georgia, 1850–1873. General Library, University of Georgia, Athens.
Minutes of the Trustees, University of Georgia, 1835–1857; *ibid.*, 1858–1871. General Library, University of Georgia, Athens. These minutes have been typed and are labeled volumes 3 and 4.
United States Census, 1860. Agriculture. Georgia, Appling-Fayette counties roll. Microfilm in General Library, University of Georgia, Athens.
United States Census, 1860. Population. Georgia, Cass-DeKalb counties roll. Microfilm in General Library, University of Georgia, Athens.

## IV.  NEWSPAPERS

Athens (Ga.) *Southern Banner,* 1855–1857.
Athens *Southern Watchman,* 1860.
*Atlanta Journal,* 1890.
Augusta (Ga.) *Weekly Constitutionalist,* 1856–1861.
Augusta *Weekly Chronicle & Sentinel,* 1847–1849, 1851.
*New York Times,* 1865.
*New-York Daily Tribune,* 1852, 1853.
*Rochester* (N.Y.) *Daily American,* 1850–1856.
*Rochester Union & Advertiser,* 1865, 1867.

## V.  PAMPHLETS

*Endowment of the Terrell Professorship of Agriculture, in the University of Georgia.* Athens: *Southern Banner* Job-Office, 1854.
*Report of President Church of the State University, to the Senatus Academicus, Held in the Senate Chamber, Thursday, November 8th, 1855.* Milledgeville, Ga.: Boughton, Nisbet & Barnes, n.d.
*University of Georgia Catalogue of the Officers and Students of Franklin College, Athens, Geo., 1854–'55.* Athens, Ga.: Reynolds & Bro., n.d.
*Ibid., 1855–1856.* Athens: Reynolds & Bro., n.d.

## VI.  PERIODICALS

*The American Agriculturist: Devoted to Improve the Planter, the Farmer, the Stock-Breeder and the Horticulturist.* New York: Vol. IX (1850).

*American Cotton Planter and the Soil of the South: Agriculture, Horticulture, Manufactures, Domestic and Mechanic Arts.* Montgomery, Ala.: Vols. I (1857)–III (1859).

*The American Farmer: Devoted to Agriculture, Horticulture, and Rural Economy.* Baltimore: Vols. for 1851, 1867–1869, 1873–1876.

*The Country Gentleman: A Journal for the Farm, the Garden and Fireside, Devoted to the Improvement of Rural Affairs, to Elevation in Mental, Moral and Social Character, and to a Record of Science, Progress, and the Times.* Albany, N.Y.: Vol. for 1858.

*The Cultivator: A Monthly Journal, Devoted to Agriculture, Horticulture, Floriculture, and to Domestic and Rural Economy. Illustrated with Engravings of Farm Houses and Farm Buildings, Improved Breeds of Cattle, Horses, Swine and Poultry, Farm Implements, Domestic Utensils, etc.* Albany, N.Y.: Vols. for 1844, 1847–1849.

*The Cultivator & The Country Gentleman: The Farm, the Garden, the Fireside. Devoted to the Science and Practice of Agriculture and Horticulture at Large, and to all the Various Departments of Rural and Domestic Economy.* Albany, N.Y.: Vols. for 1864–1872.

*Farmer and Artisan.* Athens and Atlanta, Ga.: Vols. for 1870–1871.

*The Genesee Farmer: A Monthly Journal Devoted to Agriculture & Horticulture, Domestic and Rural Economy. Illustrated with Engravings of Farm Buildings, Domestic Animals. Improved Implements, Fruits, etc.* Rochester, N.Y.: Vols. for 1845–1865. This was not a revival of a former agricultural journal under the same name.

*The Georgia University Magazine.* A student publication of the University in Athens: Vol. for 1855.

*Journal of the United States Agricultural Society.* Washington, D.C.: Vols. for 1852, 1854.

*The Plantation: Devoted to the Interests of Agriculture, Rural Economy, and the Benefit of the People.* Atlanta: Vols. for 1870–1872.

*The Soil of the South: A Monthly Journal, Devoted to Southern Agriculture and Horticulture.* Columbus, Ga.: Vols. for 1853–1856.

*The Southern Cultivator: A Monthly Journal, Devoted to the Interests of Southern Agriculture, and Designed to Improve Both the Soil and the Mind, and to Introduce a More Enlightened System of Agriculture. Illustrated with Numerous Elegant Engravings.* Augusta, Athens, and Atlanta, Ga.: Vols. for 1845–1890.

*The Southern Field and Fireside.* Augusta (later moved to Raleigh, N.C., before going out of existence): Vols. for 1859–1862.

*The Tennessee Farmer.* Nashville (not a revival of a former farm journal under the same name): Vols. for 1886–1887.

# ❧ Index ❧

Abolitionists, 83, 92
Agrarianism, 6
Agricultural and Mechanical Institute: advocated by Lee for Augusta, 29
Agricultural Association of the Planting States, 43–44
Agricultural chemistry, 7, 59, *passim*
Agricultural engineering, 62, 130–31
Agricultural schools: advocated by Lee, 4, 7–8; Western New York Agricultural School, 7; Maryland Agricultural College, 18; New York Agricultural College, 18; Agricultural and Mechanical Institute, 29; Georgia State College and Mechanical Arts, 113; College of Agriculture, University of Georgia, 113–14
Agricultural societies, 3, 16, 41, 42–43
Agriculture: defined, 7–8; scientific, 16–17, 58–59, 104, 105; diversification, 67–78; lectures by Lee on, 100–106; farming activities by Lee, 1, 35, 39, 54, 109–12, 128–35
Africa: slavery in, 80; source of slaves, 83–87; civilization in, 85, 87–88; market for Southern products, 137
Albany (N.Y.) *Country Gentleman*: Lee contributes to, 28
*Albany* (N.Y.) *Cultivator*: Lee contributes to, 21, 135
*Albany* (N.Y.) *Evening Journal*: criticizes Lee, 45
American Colonization Society, 56
Apples: for Southern crop diversification, 68; shockley, on Lee's Athens farm, 112
Athens *Farmer and Artisan*: Lee editor of, 134
Athens, Ga.: residence of James Camak, 10; visited by Lee, 12; residence of Lee, 99–115; Presbyterian Church in, 103; praised by Lee, 110
Athens *Southern Watchman*: quoted, 111
Atlanta, Ga., 29, 113

*Atlanta Journal*: quoted on Lee's death, 139
Augusta, Ga.: Lee's residence in, 11–30; milling center, 29
Augusta *Chronicle & Sentinel*: Lee agricultural editor of, 20–21, 24–25; Lee's salary on, 28; Lee resigns position on, 30
Augusta *Constitutionalist*, 83

Baer, Prof. Charles: opposed by Lee, 15
Bailey, Oscar: controversy with Lee, 111–12
Baltimore *American Farmer*: edited by John S. Skinner, 3; Lee contributes to, 135
Barbecues: described by Lee, 25
Barry, Patrick: visits World's Fair, in London, 25
Bateman, M. B.: agricultural editor, 21
Beecher, Henry Ward, 137
Belle Meade: Tennessee estate visited by Lee, 138
Bermuda grass, 74, 75
*Blackwood's Magazine*: cited by Lee on conditions in Africa, 88
Bluegrass: Lee's interest in, 75; in Greene County, 75; on University of Georgia campus, 111
*Boston Journal of Agriculture*, 44–45
Browne, D. J.: introduces sorghum to the South, 69
Browne, William Montague: professor at University of Georgia, 113
Brownlow, William G.: terrorism of condemned by Lee, 127, 132–33, 152–53 (n. 42)
Buffalo, N.Y.: Lee's residence in, 2
Buffalo (N.Y.) *Commercial Advertiser*, 3
Buffalo (N.Y.) *Honest Industry*: published by Lee, 2
*Buffalo Journal*, 3
Burke, Edmund: Commissioner of Patents, 31–32